AA002362

2006 Optics Valley of China International Symposium on Optoelectronics

Wuhan, China
November 1-4, 2006

IEEE Catalog Number: 06EX1626
ISBN: 1-4244-0816-4

Copyright © 2006 by The Institute of Electrical and Electronics Engineers, Inc.
All Rights Reserved

Copyright and Reprint Permissions: Abstracting is permitted with credit to the source. Libraries are permitted to photocopy beyond the limit of U.S. copyright law for private use of patrons those articles in this volume that carry a code at the bottom of the first page, provided the per-copy fee indicated in the code is paid through Copyright Clearance Center, 222 Rosewood Drive, Danvers, MA 01923.

For other copying, reprint or republications permission, write to IEEE Copyrights Manager, IEEE Operations Center, 445 Hoes Lane, Piscataway, New Jersey USA 08854. All rights reserved.

IEEE Catalog Number: 06EX1626

ISBN: 1-4244-0816-4

Library of Congress: 2006937434

Additional Copies of This Publication Are Available from:

IEEE Service Center
445 Hoes Lane
Piscataway, NJ 08854
IEEE Service Center
445 Hoes Lane
Piscataway, NJ 08854
Phone: (800) 678-IEEE
              (732) 981-1393
Fax: (732) 981-9667
E-mail: customer-service@ieee.org

2006 Optics Valley of China International Symposium on Optoelectronics

**Wuhan, China
1-4 November 2006**

IEEE Catalog Number: CFP0683B-POD
ISBN: 978-1-42440-816-0

Table of Contents

Waveguide Integrated Ge p-i-n Photodetectors on a Silicon-on-Insulator Platform..................1
J. F. Liu, D. Ahn, C. Y. Hong, D. Pan, S. Jongthammanurak, M. Beals, L. C. Kimerling, J. Michel

Quantum Dot Semiconductor Optical Amplifiers and Switches 5
R.V. Penty, I.H. White, S. Liu, X. Hu, M.G. Thompson, R.L. Sellin, K.A. Williams, A.R. Kovsh

Nano-optoelectronics research in WNLO.................8
Zhiping Zhoua,b, Dingshan Gaoa, Yi Wanga, Jinlin Chena, Junbo Fenga, Zhixuan Xiaa, Yao Chena

In-situ Opening Aligned Carbon Nanotubes and Applications for Device Assembly and Field Emission 12
Lingbo Zhu, Dennis W. Hess, C. P. Wong

Radio-Frequency and Millimeter-Wave Photonic Techniques for Broadband Communications and Sensor Networks..................19
Jianping Yao, Xiupu Zhang, Raman Kashyap, Ke Wu

Theoretical and Experimental Investigations of Lightcraft/ Impulsar Related Problems 26
V.V. Apollonov

Optical Buffering and Time Slot Interchanging Based on an Optical Crosspoint Switch Matrix 31
Nan Chi, Dexiu Huang, Zhuoran Wang, Siyuan Yu

Unattended Ground Sensor System Based on Fiber Optic Disk Accelerometer.................33
Yongjie Wang, Fang Li, Hao Xiao, Yuliang Liu

40Gb/S Simultaneous Inverted and Non-inverted Wavelength Conversion Based on SOA Using Transient Cross Phase Modulation..................37
Jianji Dong, Xinliang Zhang, Songnian Fu, P. Shum, Dexiu Huang

40Gb/s All-optical NOR Gate Based on Semiconductor Optical Amplifier and Fiber Delay Interferometer............41
Jing Xu, Xinliang Zhang, Deming Liu, Dexiu Huang

Sensitivity Analysis of Microring Resonator Based Biosensor: the Quality Factor Perspective 44
Zhixuan Xiaa, Zhiping Zhoua

All-optical RZ to NRZ Format Conversion with Tunable Fiber Based Delay Interferometer 47
Yu Yu, Xinliang Zhang, Dexiu Huang

Study of Coupled-Resonator-Induced Transparency in 3×3 Coupler Based Dual Microring Resonators.................51
Xiaobei Zhang, Dexiu Huang, Xinliang Zhang

Photonic Crystal Taper for Efficient Coupling and Smooth Mode Profile Conversion 54
Jing Liua, Dingshan Gaoa, Zhiping Zhou

Tunable Wavelength Multicasting Using Pulsed Pumping Four-Wave Mixing in a Highly Nonlinear Fiber.............57
Jian Wang, Junqiang Sun, Qizhen Sun, Hui Cao

FDTD Simulation of Band Structure and Mode Distribution for Plasmonic Crystals.................60
Bin Zhanga, Yi Wanga, Zhiping Zhou

Measurement of Gain Curves for Semiconductor

Optical Amplifier Utilizing Hakki-Paoli Method With Wavelet Denoise and Deconvolution Process 63
Lei Liu, Xinliang Zhang, Dexiu Huang

OFDM-ROF System and Performance Analysis of Signal Transmission.................67
Linghui RAO, Xiaoqiang SUN, Wei LI, Dexiu HUANG

iv

Waveguide Integrated Ge p-i-n Photodetectors on a Silicon-on-Insulator Platform

J. F. Liu, D. Ahn, C. Y. Hong, D. Pan, S. Jongthammanurak, M. Beals, L. C. Kimerling, and J. Michel
Department of Materials Science and Engineering, Massachusetts Institute of Technology, Cambridge, MA 02139, USA
A. T. Pomerene, D. Carothers, C. Hill, and M. Jaso
BAE Systems, Semiconductor Technology Center, Manassas, VA 20110, USA
K. Y. Tu, Y. K. Chen, S. Patel, M. Rasras, D. M. Gill, and A. E. White
Lucent Technologies Bell Laboratories, 600 Mountain Avenue, Murray Hill, New Jersey 07974, USA

Abstract—We present selectively grown Ge p-i-n photodetectors coupled to high index contrast Si(core)/SiO₂(cladding) waveguides on a silicon-on-insulator (SOI) platform. Two coupling schemes, namely butt-coupling and vertical coupling, were demonstrated in this study. With the butt-coupling scheme we have achieved a high responsivity of 1.0 A/W at 1520 nm and a 3dB bandwidth greater than 4.5 GHz at 1550 nm. With the vertical coupling scheme, where the light couples from a Si waveguide evanescently to the Ge detector on top of it, a responsivity of 0.22A/W and a 3dB bandwidth of ~1.5 GHz have been demonstrated at 1550 nm. The devices were fabricated on a standard 180 nm industrial complementary metal oxide semiconductor production (CMOS) line, and can be integrated with CMOS circuitry for electronic and photonic integrated circuits.

I. INTRODUCTION

Electronic and photonic integration has been envisioned as promising solution to the interconnect bottleneck on-chip [1]. An electronic-photonic integrated circuit (EPIC) on silicon platform requires waveguide-integrated photodetectors to receive optical signals through the input waveguides and convert them to electrical signals. Despite of their high performance at telecom wavelengths, InP based photodetectors are difficult to integrate with Si-based electronic circuits. In recent years Ge has emerged as an alternative detector material with excellent responsivity, high bandwidth and the potential of monolithic integration on silicon [2-7]. In this paper we present high performance Ge p-i-n photodetectors integrated with high index contrast Si(core)/SiO₂(cladding) waveguides on a SOI platform. Two coupling schemes are demonstrated: butt-coupling and vertical coupling. In the butt-coupling scheme the output end of the Si waveguide directly inputs to the Ge photodetector, while in the vertical coupling scheme the light couples evanescently from the Si waveguide to the Ge detector on top of it. The devices were fabricated completely on a standard 180 nm industrial CMOS production line, and can be integrated with CMOS circuitry to achieve electronic and photonic integration on Si.

II. OPTOELECTRONIC MEASUREMENT OF WAVEGUIDE-INTEGRATED PHOTODETECTORS

Fig. 1 schematically shows the measurement setup for waveguide-integrated photodetectors. The performance of the photodetector was measured by coupling light into the waveguide and monitoring the electrical response of the detector. To measure the responsivity, light is coupled through an optical fiber to the waveguide and the photocurrent, I_{ph}, is measured from the detector. The optical power input to the photodetector, P_{in}, is determined from the measured transmitted optical power of a reference waveguide, $P_{out,ref}$, and the measured transmission loss γ (in unit of dB per unit length) in the waveguide. The responsivity R is given by

$$R = I_{ph}/(10^{\gamma L/10} P_{out,ref}) \qquad (1)$$

where L is the distance from the input point of the detector to the output end of the waveguide, as shown in Fig. 1.

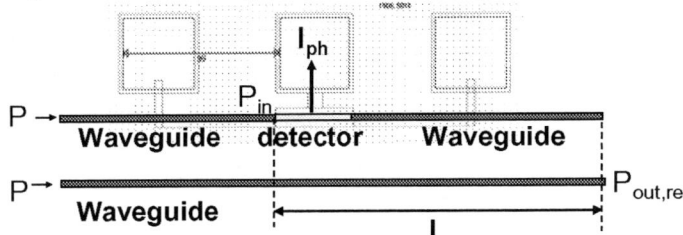

Fig. 1 Schematic diagram showing the optoelectronic measurement of a waveguide-coupled photodetector.

The bandwidth of the photodetectors was measured either directly in the frequency domain by using a 10 GHz optical modulator, a 50 GHz network analyzer, and a trans-impedance amplifier (TIA), or in the time domain by inputting a 1 ps-long laser pulse into the waveguide and record the response of the detector with a 50 GHz digital oscilloscope. In the latter case, a Fourier transform was used to convert the data to frequency domain and estimate the 3dB bandwidth.

III. GE PETECTORS BUTT-COUPLED TO SI WAVEGUIDES

A. Fabrication process

The fabrication process of butt-coupled Ge photodetectors is schematically shown in Fig. 2. Butt-coupling from amorphous silicon waveguides to Ge photodetectors was achieved by growing Ge selectively to fill trenches opened through a SiO₂ layer, followed by a chemical mechanical polishing (CMP) process to planarize the top of the selectively grown Ge. The Ge material was alloyed with 0.8 at. % Si to achieve

1-4244-0816-4/06/$25.00 ©2006 IEEE

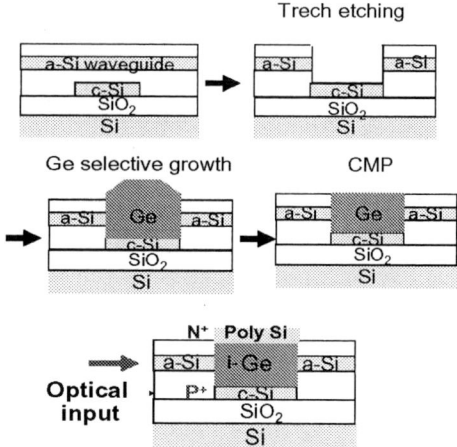

Fig. 2. Schematic diagram of the process flow to achieve butt-coupling from Si waveguide to Ge photodetector.

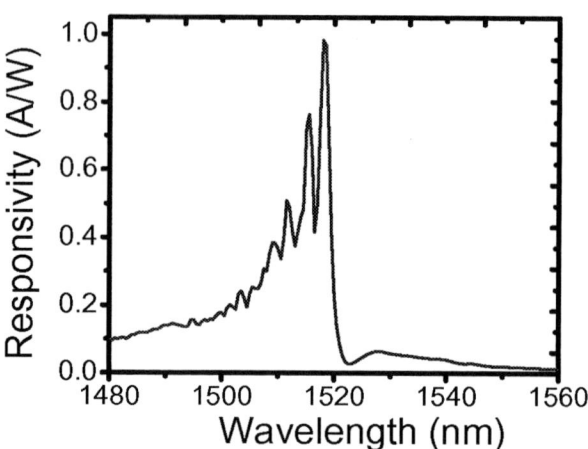

Fig. 4. Spectral responsivity of a 50 μm-long, butt-coupled Ge photodetector at 3 V reverse bias.

modulator/detector integration, and the details will be reported elsewhere [8]. Fig. 3a shows a cross-sectional scanning electron microscopy (SEM) image of a trench over-filled with Ge. The Ge overgrows the SiO_2 after completely filling the trench, forming (111) and (311) facets in the extruding parts. A CMP process was employed to remove the overgrown Ge. Fig. 3b shows the atomic force microscopy (AFM) image of a Ge-filled trench after CMP. The top of the Ge stripe in the trench is relatively smooth after CMP, with a RMS roughness of ~2nm. The crystalline Si layer of the SOI wafer was implanted with boron before the Ge growth to form the p-type electrode of the p-i-n diode, while an n^+ poly Si layer on top of Ge forms the n-type electrode. The Si waveguide is 500 nm in width and 200 nm in height to achieve single mode at 1550 nm. The width of the butt-coupled detector is 600 nm and the height is also 600 nm. The coupling efficiency is estimated to be 86% from simulations.

can be avoided by introducing a bend in the Si waveguide before the light couples to the detector to filter out higher order modes.

Fig. 5 shows the frequency response of the photodetector at 1550 nm with 3V reverse bias. The sharp roll-off at 4.5 GHz is due to the bandwidth limitation of the TIA instead of the detector itself. Therefore, the bandwidth of the detector is >4.5 GHz. With a responsivity of 1.0 A/W and a bandwidth greater than 4 GHz around 1550nm, the device has promising applications in electronic and photonic integrated circuits.

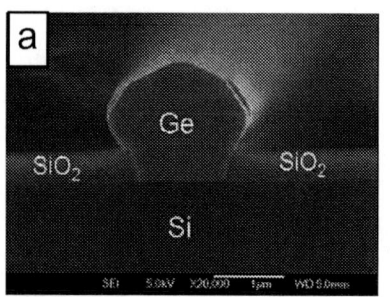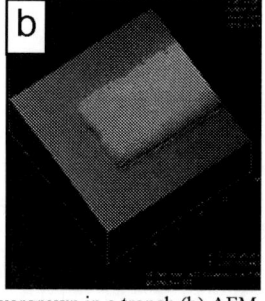

Fig. 3. (a) Cross-sectional SEM image of Ge overgrown in a trench (b) AFM picture after removing the overgrown Ge by CMP.

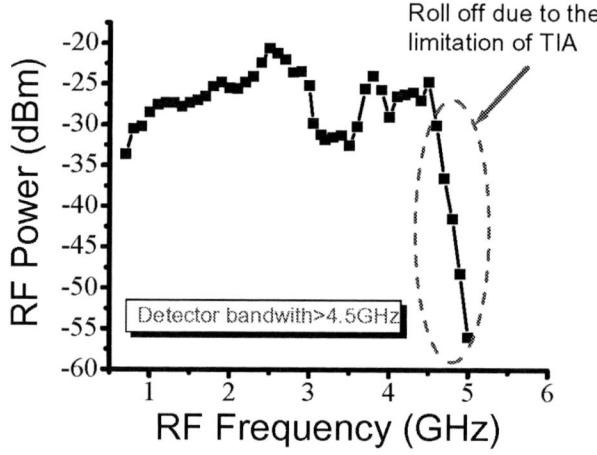

Fig. 5. Frequency response of a 50 μm-long, butt-coupled Ge photodetector at 3V reverse bias.

B. Device performance

Fig. 4 shows the spectral responsivity of a 50 μm-long device at 3V reverse bias. With 0.8% Si the direct band edge of Ge is shifted to ~1520 nm. Therefore, a significant increase in the responsivity was observed as the wavelength decreases to 1520 nm. At 1520 nm the responsivity is ~1.0 A/W, and the external quantum efficiency is 83%. This is close to the result of theoretical modeling (86%). The oscillations at wavelengths <1518 nm are due to the coupling to second order modes, and

IV. VERTICALLY COUPLED GE PHOTODETECTORS

A. Fabrication process

Fig. 6 schematically shows the fabrication process of the vertically coupled Ge photodetectors. First, we patterned and implanted boron into the crystal Si layer of the SOI only in the areas in which Ge photodetector were to be grown later (Fig. 6a). Crystalline Si waveguides with taper structures were defined by lithography such that the width of the Si bus waveguide reduced to <50 Ω.

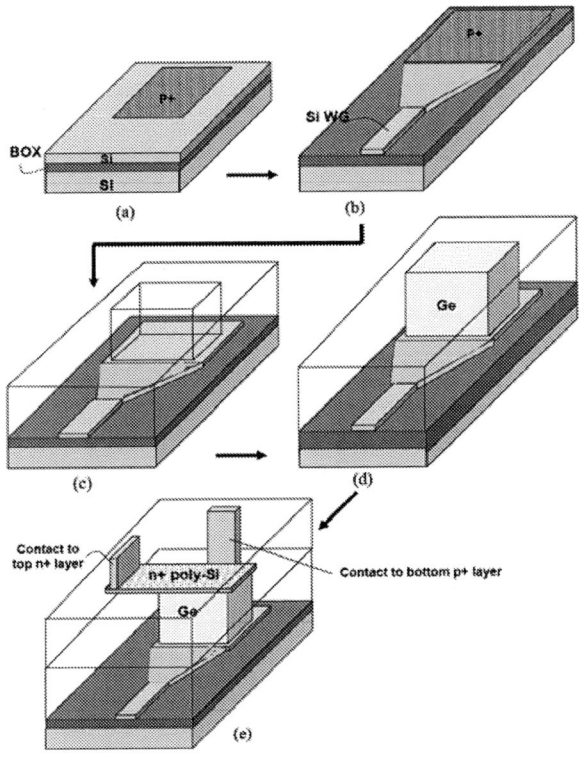

Fig. 6. Schematic diagram of the process flow to fabricate vertically-coupled Ge photodetectors.

gradually increases to that of the vertically coupled Ge photodetector (2.5μm) (Fig. 6b). An oxide layer was deposited and trenches were etched to expose the p$^+$ Si at the bottom (Fig. 6c). Like butt-coupled detectors, Ge with 0.8 at. % Si was selectively grown in the trenches on top of p$^+$ Si and planarized by CMP (Fig. 6d). An n$^+$ poly Si layer on top of Ge forms the n-type electrode of the p-i-n diode structure (Fig. 6e). In this structure, light travels in the crystalline Si waveguide and couples into the Ge detector by evanescent coupling.

B. Device performance

Fig. 7 shows the spectral responsivity of a 5μm-long and a 20 μm-long vertically-coupled Ge detector. Since the direct band gap of the Ge$_{0.992}$Si$_{0.008}$ material is at 1520 nm, the 5μm-long device shows a roll-off in the responsivity as the wavelength increases above 1520 nm. Nevertheless, a higher responsivity at longer wavelength is achieved with a 20 μm-long device because the light gets fully absorbed with long enough detector length even the absorption coefficient is low. A responsivity of 0.22 A/W was achieved around 1550 nm with a 20 μm-long device.

Fig. 8 shows the temporal response of a 5μm-long, vertically coupled Ge detector to a 1 ps-long pulse centered around 1550 nm. The Fourier transform of the data reveals a 3dB bandwidth of ~1.5 GHz, mainly limited by the RC delay. The cause of the RC delay limit is the relatively high series resistance of the device due to fabrication errors. A much faster device with a bandwidth of >20 GHz can be achieved if the series resistance is

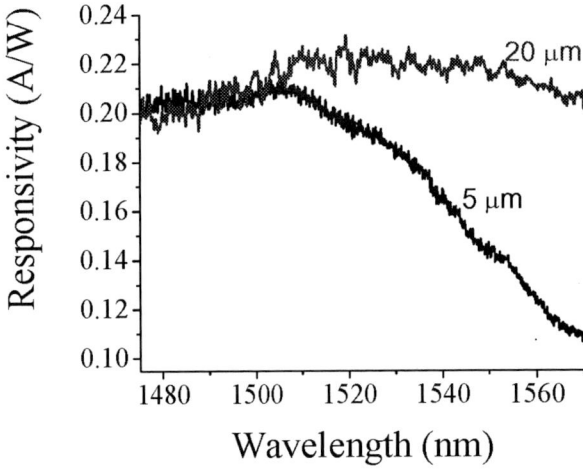

Fig. 7. Spectral responsivity of a 5μm-long and a 20 μm-long vertically-coupled Ge photodetector.

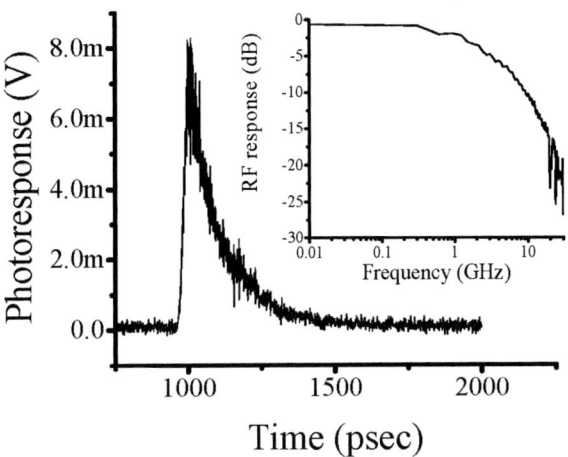

Fig. 8. (a) Impulse response of a 5μm-long, vertically-coupled Ge photo detector at 4V reverse bias to a 1-ps long pulse centered around 1550 nm, The inset shows the Fourier transform of the data, revealing a 3dB bandwidth of ~1.5 GHz.

V. CONCLUSIONS

We have demonstrated waveguide-integrated Ge photodetectors on a SOI platform with two different coupling schemes: butt-coupling and vertical coupling. Both schemes have achieved efficient coupling from the Si waveguide to the Ge detector, and the 3-dB bandwidth of the detectors is in the order of GHz. These two coupling schemes can be applied to different cases depending on the circuit architecture and device fabrication process. The devices were fabricated on a standard 180 nm industrial CMOS production line, and can be integrated with CMOS circuitry to achieve electronic and photonic integration on Si.

Acknowledgment

This work was sponsored under the Defense Advanced Research Projects Agency's (DARPA) EPIC program supervised by Dr. Jagdeep Shah. The program is executed by the Microsystems Technology Office (MTO) under Contract No. HR0011-05-C-0027. The authors would like to thank Dr. John Yasaitis for helpful discussions.

References

[1] International Technology Roadmap for Semiconductors (ITRS), Interconnect Chapter, 2005 edition,

[2] L. Colace , G. Masini, G. Assanto, H. C. Luan, K. Wada, L. C. Kimerling, *Appl. Phys. Lett.* **76**, 1231 (2000)

[3] S. Fama, L. Colace, G. Masini, G. Assanto, and H.-C. Luan, *Appl. Phys. Lett.* **81**, 586 (2002)

[4] S. J. Koester, J. D. Schuab, G. Dehlinger, J. O. Chu, Q. C. Ouyang, and A. Grill, Session V.A-4, 62nd Annual Device Research Conference, Notre Dame University, Notre Dame, Indiana, Jun 22, 2004.

[5] O. I. Dosunmu, D. D. Cannon, M. K. Emsley, L. C. Kimerling, and M. S. Unlu, *IEEE. Photon. Technol. Lett.* **17**, 175 (2005)

[6] Z. H. Huang, J. Oh, and J. C. Campbell, *Appl. Phys. Lett.* **85**, 3286 (2004)

[7] J. F. Liu, J. Michel, W. Giziewicz, D. Pan, D. D. Cannon, D. T. Danielson, S. Jongthammanurak, K. Wada, and. C. Kimerling, *Appl. Phys. Lett.* **87**, 103501 (2005)

[8] J. F. Liu, D. Pan, S. Jongthammanurak, K. Wada, L. C. Kimerling and J. Michel, submitted to *Optics Express*

Quantum Dot Semiconductor Optical Amplifiers and Switches

R.V. Penty[1], I.H. White[1], S. Liu[1], X. Hu[1], M.G. Thompson[1], R.L. Sellin[1], K.A. Williams[1], A.R. Kovsh[2]

1: University of Cambridge, 9 JJ Thompson Avenue, Cambridge, CB3 0FA, United Kingdom
2: NL Nanosemiconductor GmbH, Konrad-Adenauer-Allee 11, 44263 Dortmund, Germany

Abstract—**We report a study of the cascadability of QD-SOAs in a recirculating loop, showing that it is possible to cascade 15 SOAs before error free operation is lost. Furthermore, a monolithically integrated QD-SOA based add-drop switch is reported which exhibits very low (<0.1dB) path dependent switching power penalty.**

I. INTRODUCTION

Quantum dot semiconductor optical amplifiers (QD-SOAs), with their broad gain spectrum, ultra-fast gain recovery rates and consequent potential for low distortion are emerging as high performance devices [1] for prospective applications such as in power boosters, in-line repeaters [2], and reconfigurable network switches [3]. In several of these applications, it is important to be able to cascade multiple amplifiers, while ensuring that penalties accrued from signal distortion and noise are tolerably low after the cascade. Work to date has however focussed in the main on the stand-alone performance of single components. Studies to date on quantum well SOAs, operating in the 1.55μm wavelength band, have been performed for up to eight cascaded amplifiers [4]. This work therefore describes a study of the cascaded performance of QD SOAs and describes what is believed to be the first monolithically integrated QD-SOA based add-drop switch.

II. QD SOA DEVICE AND CASCADE EXPERIMENT

The quantum dot amplifier used in the first part of this work is grown by molecular beam epitaxy on a GaAs substrate with 15 layers of InGaAs QDs. The waveguide width is 4μm and the length of the amplifier is 4mm. The device is deeply etched to ensure strong optical confinement. It operates at relatively low current with only 70mA being required to provide 12dB fibre-to-fibre gain at the peak wavelength 1283nm. At this operating point, the in-fibre noise figure as evaluated from a direct comparison of input and output signal to noise ratios is 6dB. The corresponding in-fibre saturation output power is estimated to be 0dBm.

To assess cascadability, the QD SOA is placed inside a recirculating loop system as shown in figure 1. Here a 3dB fibre coupler is used to load and unload the recirculating loop. The input transmitter on the left of the figure is optically gated using a QW semiconductor optical amplifier which also serves as a power booster for the tunable laser. The 10Gb/s data is imprinted on the optical signal by a Mach-Zehnder modulator which has been optimised for operation at 1.55μm. This

therefore leads to a compromised extinction ratio as shown as an inset in figure 2. The transmitter therefore provides a 10Gb/s non-return-to-zero pseudo-random binary sequence, for a duration of approximately 50ns at a wavelength of 1290.8nm. The data, gated by the input SOA gate, is then loaded to the recirculating loop once per test period. The loop is cleared by gating the current of QD SOA under test itself. The QD SOA is the only active element within the loop and therefore the only source of noise and distortion which will account for degradation between consecutive loops. It is worth noting that no filter has been placed within the loop to remove ASE.

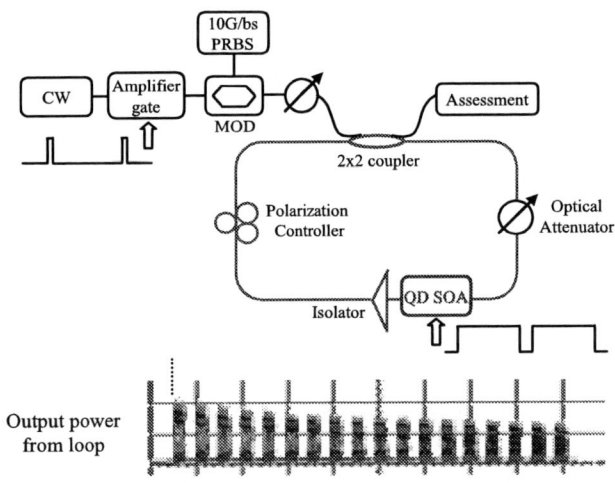

Fig 1: A schematic diagram for the experimental assessment of the QD SOA (top) with an output time trace from the loop showing recirculations (bottom)

III. RESULTS AND DISCUSSION

The QD amplifier is gated with a peak current of 70mA and here has a net fibre-fibre gain of 9.4 dB. The total loop loss when the QD amplifier is bypassed is tuned to 9.5 dB by introducing 3.5 dB loss with the variable optical attenuator. The loop circulation time is 100ns, and so within the test period of 1792 ns it is possible to assess circulations of up to 15 cascades. The input signal to the QD SOA is optimized to be -11.6 dBm. The assessment scheme comprises an optical filter with bandwidth of 0.3 nm, a variable optical attenuator to enable penalty assessment and an optically pre-amplified receiver. Eye diagrams are recorded at the output of the loop for each iteration around the loop. A quantitative assessment is made by evaluating the Q factor using the in-built function of a digital communications analyzer. The back to back performance is recorded as having a Q factor of 9.3. Figure 2 shows a small degradation for each circulation. A maximum number of 11 cascaded amplifications is observed while the Q-factor is maintained at greater than 6, equivalent to a BER of < 10^{-9}.

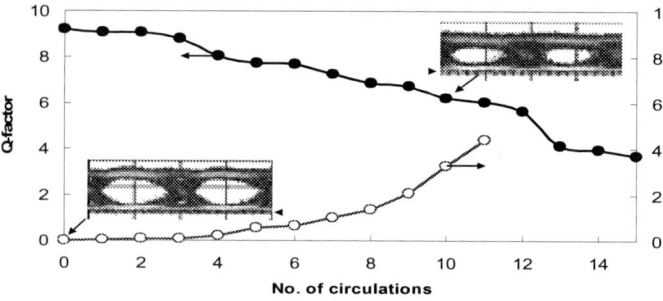

Figure 2: Q factor and estimated power penalty for consecutive circulations around the loop

The power penalty is estimated by setting the variable optical attenuator for a Q factor of 6 and measuring the required loss to achieve this. The optical signal is believed to be degraded primarily by the amplified spontaneous emission noise. As a result, the signal becomes increasingly noisy for consecutive loops. However seven amplifiers can be cascaded with an estimated power penalty of less than 1 dB. This is believed to be a record for a QD SOA.

Cascaded operation is demonstrated for the first time in QD-SOAs with up to 11 circulations being possible for a Q of greater than 6. This low penalty, filter free, cascaded performance of QD SOA based devices is thus envisaged to allow for a broad range of applications in reconfigurable networking and transparent data networking solutions.

IV. MONOLITHIC QD-SOA ADD-DROP SWITCH

To date research on QD-SOAs has mainly focused on discrete components. Here we extend the work on cascaded QD-SOA operation described above and report what is to our knowledge the first monolithic 1.3 μm InGaAs QD add-drop switch and hence what represents a step function in QD device integration.

The switch active layer consists of a 10 layer stack of InGaAs QD layers in a GaAs slab waveguide. Two ridge waveguides, each comprising three SOAs, are defined from left to right at both the top and bottom of the chip shown in fig. 3. Two interconnecting waveguides lie vertically between them. A total of eight interconnected SOAs are thus separately addressed on the chip, four serving as input and output guides, and four as gates. Each of four input ridge waveguides width is 3 μm and this is expanded linearly to a 6 μm width over a length of 150 μm to the splitter before connecting to perpendicular SOA gates.

The splitter SOAs are always operated in the on state whilst the gate SOAs (sections 2, 4 5 and 7 in fig. 3) are switched. The splitters comprise 45° totally internal reflecting mirrors to route the light to perpendicular and through paths. The totally internally reflecting mirrors are formed by focused ion beam etching vertical slots through the active layer (see fig. 3 inset). Patterned p-type electrodes allow individual addressing of the

amplifier gates and splitters. The isolation between the electrodes exceeds 5 kΩ for all electrode combinations. The drop, add and through paths are enabled by inputting current to section 4, 7 and 5 SOA gates respectively. Enabling the section 2 gate allows the add port to connect directly to the drop port. The overall chip area is only 2.55x0.85 mm^2.

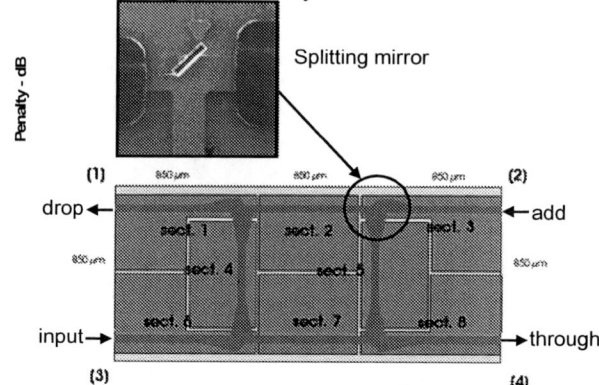

Figure 3. Schematic of the switch structure, showing the waveguides and the SOA contact layout,, with a secondary electron image of a mirror at the end of the waveguide splitter shown as an inset. The arrows indicate the input and output waveguides.

V. OPERATING CHARACTERISTICS

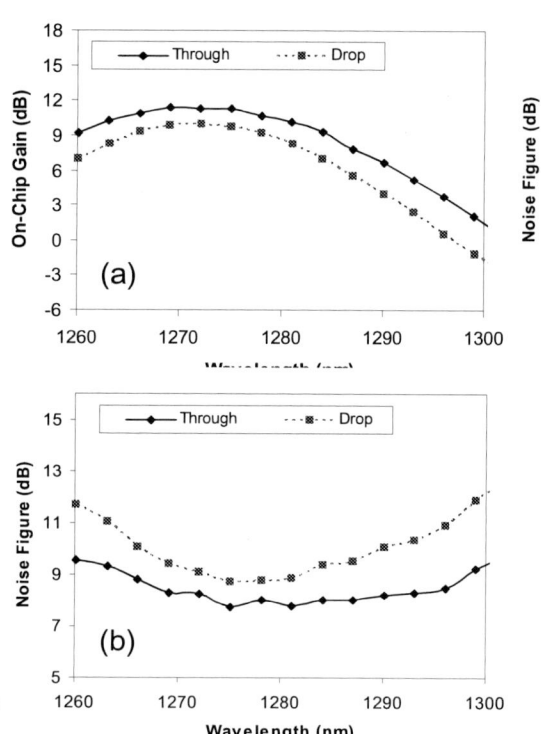

Figure 4. On-chip gain (a) and fibre couled noise figure (b) as a function of wavelength and path through the switch at a bias current of 70 mA per gate.

The gain characteristics and noise performance are assessed using a -15 dBm continuous wave tuneable laser which couples light via a fibre lens into the device. The coupling loss between each of the input and output fibre lenses and the chip is estimated from the measured photo current to be 3.1 dB. The light is then split between the through and perpendicular

6

directions by the splitter and the corresponding amplifier gates. The input and output waveguide splitters are biased at 70 mA. The net on-chip gain, inclusive of splitter losses, is 10 dB from the input port to the drop port and 11 dB for the through path at the peak gain wavelength of 1272 nm, as shown in fig 4 (a). This leads to an estimate of the splitting ratio of 0.48/0.52, assuming equal gains in each SOA. The slight mismatch in path gain can be easily balanced by fine tuning SOA currents. Whilst the chip exhibits high polarisation dependent gain, much progress in being made in achieving polarisation independent operation in QD SOA devices [5].

The device noise figure is measured by a direct comparison of optical signal-to-noise ratio before and after the chip. The lowest - noise figure is at a wavelength of 1275 nm and is 7.7dB for the through path and 8.7dB for the drop path. Wavelength dependent noise figures are shown in Figure 4 (b). It should be noted that the on-chip noise figure is lower by the 3.1dB output coupling loss and better coupling, e.g. via expanded mode waveguides, would reduce the system noise figure accordingly.

VI. DYNAMIC OPERATION

A 10Gb/s 2^{31}-1 pseudo random binary sequence from a pulse pattern generator (PPG) was used as the input data to a Mach-Zehnder modulator. The data was routed through the switch by driving the gate SOAs located between the input and the through port (section 8) and the input and the drop port (section 1) with complementary electrical waveforms with a 250 μs period. The outputs from the through and drop ports are recorded using a digital communication analyser (DCA), the resulting waveforms being shown in fig. 5. As can be seen, the switch is successfully routing the data between the through and drop ports with little evidence of crosstalk.

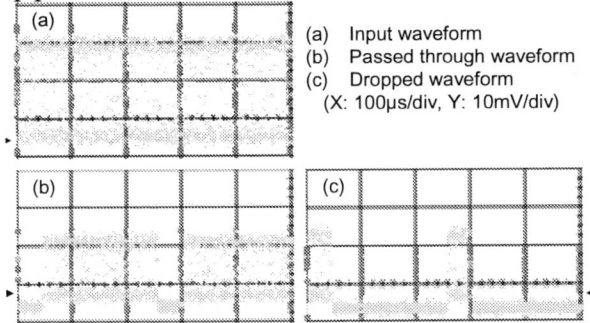

(a)   Input waveform
(b)   Passed through waveform
(c)   Dropped waveform
        (X: 100μs/div, Y: 10mV/div)

Figure 5. Switching of 10Gb/s 2^{31}-1 PRBS data.

The performance of the switch is further assessed with a bit error rate measurement for paths through the switch using a 70 mA on state current and 0 mA off state current for each gate. The average in-fibre input signal power launched into the switch is -10 dBm. The output signal from the QD switch is then passed through a 0.3 nm optical bandpass filter before it is further amplified by a commercial SOA working within the linear regime before being input to a 10 Gb/s receiver circuit. The eye diagrams with and without the switch at a receiver power of -21.5 dBm are recorded using a DCA and the corresponding Q-factors are measured to be 8.6 and 8.1, respectively, as shown in fig. 6. As can be seen, they exhibit no observable patterning.

Power penalties are obtained through bit error rate measurement with and without the inclusion of the QD SOA switch. 10^{-9} bit error rate power penalties of 0 dB, 0.1 dB and 0.1 dB may be estimated from fig 4 for the through, add and drop ports respectively.

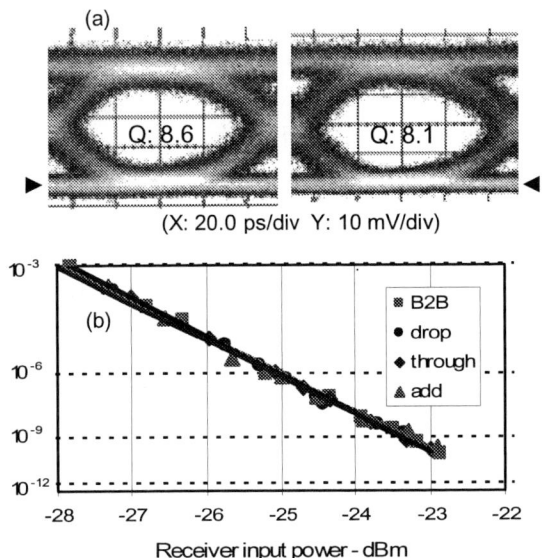

Figure 6. Input (left) and received (right) eyes (a). and bit error rate as a function of received mean packet power for through, drop and add paths (b).

VII. CONCLUSIONS

The first monolithic 2x2 QD SOA based space switch has been designed, fabricated and demonstrated. On chip gains of 10dB, with on chip noise figures of 4.6 – 5.6 dB are measured. Minimal patterning is observed, unlike with the equivalent quantum well device, resulting in very low (<0.1dB) switching penalties for the switch operated in add-drop mode. These excellent results show the strong potential for QD SOA based monolithic switches for high performance photonic switching applications.

REFERENCES

[1] T. Akiyama et al, "Quantum Dots for Semiconductor Optical Amplifiers", OFC '05 (2005)
[2] D.R. Zimmerman, L.H. Spiekman, "Amplifiers for the Masses: EDFA, EDWA, and SOA Amplets for Metro and Access Applications", Journal of Lightwave Technology 22 (1): 63-70 JAN 2004
[3] K.A. Williams et al, "Integrated Optical 2x2 Switch for Wavelength Muliplexed Interconnects", IEEE Journal of Selected Topics in Quantum Electronics, Vol. 11, No. 1, Jan 2005
[4] L. H. Spiekman et al, "8x10 Gb/s DWDM Transmission over 240 km of Standard Fiber Using a Cascade of Semiconductor Optical Amplifiers", Photonics Technology Letters, Vol 12, No. 8, 2000
[5] M Sugawara et al J. Phys. D: Appl. Phys. Vol. 38, pp. 2126-2134, 2005

Nano-optoelectronics research in WNLO

Zhiping Zhou[a,b], Dingshan Gao[a], Yi Wang[a], Jinlin Chen[a], Junbo Feng[a], Zhixuan Xia[a], Yao Chen[a]

[a]Wuhan National Laboratory for Optoelectronics, Huazhong University of Science and Technology,
Wuhan, Hubei 430074, China
[b]School of Electrical and Computer Engineering, Georgia Institute of Technology,
Atlanta, Georgia 30332, USA

Abstract—The research progress in nano optoelectronic devices and their integration at Wuhan National Laboratory for Optoelectronics are summarized. Integrations on different material platforms are described, but emphasis is given to new micro/nano scale emitters, detectors, and light beam controlling devices. The perspective of micro/nano scale monolithic integration of optical devices and electronic devices on a single chip by standard CMOS technologies is presented. The possibility of using these devices for communications, optical interconnections and bio-sensing, is also discussed.

I. INTRODUCTION

Nanoscience and nanotechnology have been developed quickly in all aspects of the human society in recent 10 years. The price, performance, and reliability that we have come to expect from today's miniaturized electronic circuits, information storage devices, and chemical/biological sensors are due in many cases to embedded nanoscale materials and structures. Driven by this desire, U.S., E.U., and other countries have been investing huge amounts of money and resources in the development and integration of micro/nano devices and systems.

Optoelectronic devices impact many areas of society, from simple household appliances and multimedia systems to communications, computing, and medical instruments. Intelligence, integration, low cost, and high reliability are becoming the main characteristics for next generation optical communication systems. At the same time, with the rapid development of high performance computers, CPU works faster and memory stores more, hence high density data communication within the computer systems becomes the bottleneck and optical interconnection becomes a very important subject for data exchange within computer and even within chip. Given the demand for ever more compact and powerful systems, there is growing interest in the development of nanoscale devices that could enable new functions and/or greatly enhanced performance [1]. Using tools originally developed for silicon integrated circuit industry, people are now fabricating nanoscale photonic devices, such as emitters, detectors, and photonic crystal based devices and waveguides, from silicon and other materials, in hopes of integrating optical and logic circuits on a single silicon chip. Doing so would enhance the performance of the high density data communications in optical communication systems and computer communication systems and reduce system size, power and cost.

This paper summarizes the research progress in nano optoelectronic devices and their integration for communications and bio-sensing at Wuhan National Laboratory for Optoelectronics (WNLO). Emphasis will be given to new micro/nano scale devices such as Silicon based emitters, detectors, light beam control devices, etc. The definition of the "Silicon Based" means that the optoelectronic devices are fabricated using silicon or other materials grown on silicon substrates.

II. SILICON BASED EMITTER

Silicon materials are widely used by standard microelectronics facilities which have shown great impact on the life of human beings. However, the growing needs towards low cost and high speed devices require the recombination of electronics and photonics within a same silicon chip. Several silicon photonics devices have been demonstrated using standard CMOS process, such as silicon photonic-crystal waveguide [2] and silicon optical modulator [3] etc. Nevertheless, there is still lacking of an efficient silicon source. In the following, an erbium-doped silicon emitter, with the help of surface plasmon polarions (SPPs), is introduced [4].

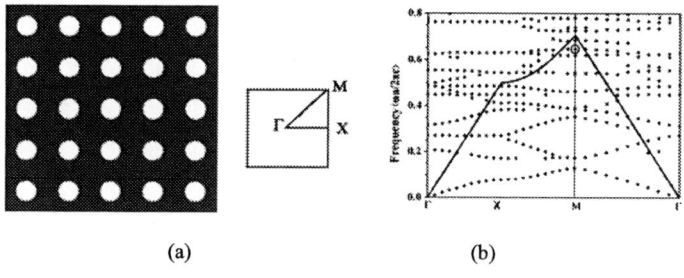

(a)   (b)

Fig. 1. (a) The two dimensional plasmonic crystal with nanorods in square array embedded on silicon, gray for silicon, white for silver, and the wavevector path of 1st Brillouin Zone for square lattice. (b) Band structure of the two dimensional plasmonic crystal for $f=0.54$. The solid line stands for the light line in air. The point within circle presents the band edge at M point.

The scheme of the silicon light emitter is shown in Figure 1. The two dimensional plasmonic crystal with nanorods in square array is embedded on silicon and the erbium ions are doped

1-4244-0816-4/06/$25.00 ©2006 IEEE

within the silicon material. When the light is incident on the structure, the erbium ions are excited. When the separation between the erbium ions and the SPPs are in the range of micrometers, the energy of erbium ions can be partial transferred into the SPPs, which are bounded on the interface between the silver and silicon. As it is shown in Figure 1 (b), by choosing the filling factor as 0.54, the energy of SPPs is localized at the edge of the band gap which is corresponding to the wavelength of 1550nm. Because of the mismatch between the SPPs and photons, the energy of SPPs can't be coupled out directly. After adding or subtracting a multiple of external wavevector, provided by the two dimensional plasmonic crystal, the energy of the SPPs can be reradiated into far field. The emission efficiency of the erbium ions at the wavelength of 1550nm by using two dimensional plasmonic crystal can increase nearly 40 times comparing to that of one dimensional crystal.

III. SILICON BASED DETECTORS

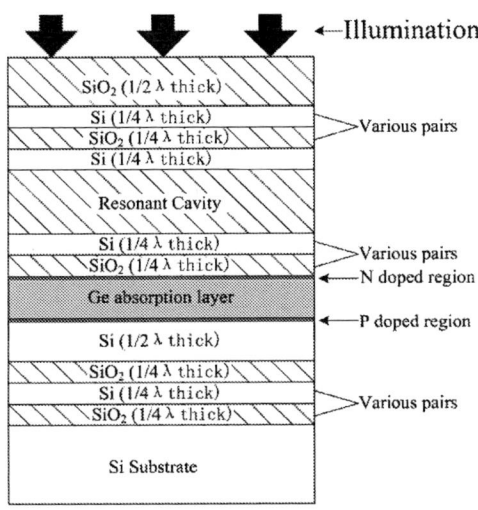

Fig.2 Schematic of the narrowband Ge-on-Si RCM-RCE photodetector

Silicon based detectors are very promising to be used in optical communications and interconnections, which have the indispensable potential to be integrated with silicon based electronic devices and will offer a much cheap way to advance optical communications and computer data communications.

We have demonstrated a structure called resonant cavity mirror-resonant cavity enhanced (RCM-RCE) for ultranarrow band and high efficiency Ge-on-Si detectors [5], which is shown in Figure 2. Based on a detailed comprehensive simulation, the proposed structure has been shown to have an ultranarrow band (<1 nm), which is much narrower than the reported best result of 5 nm and also narrower than a separate FP filter with the same reflectance profile in front of a conventional detector, and a very high quantum efficiency of near unity photoresponse. In general, the proposed structure could be specifically designed and simulated for different applications and realized by CMOS compatible technologies.

IV. SILICON BASED LIGHT CONTROLLING DEVICES

A. A. Silicon based photonic crystal modulator

Photonic crystals (PC) are seen as a way to ultracompact optoelectronics where the light emission and propagation properties will be fully controlled in micro/nano size by the design of the structures.

In the field of silicon modulator, conventional silicon Mach-Zehnder (MZ) modulators are based on rib waveguides, which usually need one-half to several millimeters to achieve the required phase shift in MZ structures. Photonic crystals waveguide, which has much larger dispersion than ordinary waveguide, can provide great phase shift even in very short length. If it is used as the modulating arm of the MZ, the length and power consumption of the MZ modulator could be reduced by magnitudes [6].

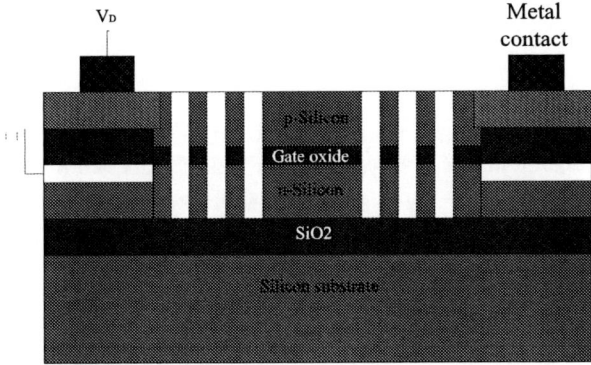

Fig.3 Section view of a MOS based PC-MZ modulator

The weakness of silicon based modulator is its low modulation speed. We have designed a MOS based PC-MZ modulator, which is shown in Figure 3. In this structure, the carrier lifetime can be shorten to avoid the speed delay caused by carrier recombination. We believe the modulation speed of our PC-MZ modulator should exceed 3GHz in 2 years.

B. Polarizing beam splitter

Polarizing beam splitters (PBS) are very important components for many applications, such as polarization-based imaging systems, magneto-optic data storage in optical information processing and optical switching, routing and isolating in optical communication. These applications require that a PBS provide high extinction ratios, tolerate a wide angular bandwidth, have a broad wavelength range of incident waves, and be compact for efficient packaging.

We developed a novel metal-grid PBS using a nanofabrication technique [7], as shown in Fig. 4, which embedded the metal grating under a homogeneous material as the grating. This kind of embedded metalgrid PBS can provide other beneficial optical properties and can protect the polarizer. The PBS not only provides high polarization efficiency for two orthogonal polarizations but also can be operated with an optical signal of wide angular tolerance and broad wavelength range. We used the effective-medium theory for the initial design and software based on rigorous coupled-wave analysis for optimization of the PBS. High polarization efficiency and low insertion loss with a broad wavelength range (900–1700nm) and a wide angular tolerance are obtained by optimization of the designed structure.

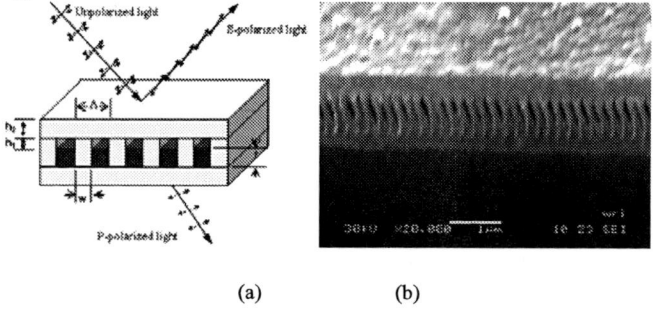

(a)   (b)

Fig. 4 (a) Schematic of an embedded metal-wire nanograting. p-polarized light is transmitted and s-polarized light is reflected. (b) Scanning electron microscope photograph of the embedded metal-wire nanograting with a period of 200 nm and a duty cycle of 0.75.

C. Binary blazed grating coupler

Coupling light from fiber (or "free space") to nano-waveguide is a very important task because an integrated circuit is useless without an interface to the outside world. But the small size of nano-waveguide makes the coupling more difficult, resulting from huge mode mismatch and strict alignment.

Here, a binary blazed grating coupler that allows surface vertical coupling is proposed as a promising component for compact high efficient waveguide coupling. The binary blazed grating is composed of subwavelength pillars with uniform height, which can be easily fabricated by one etching step. It is found that this element substantially outperforms standard grating couplers.

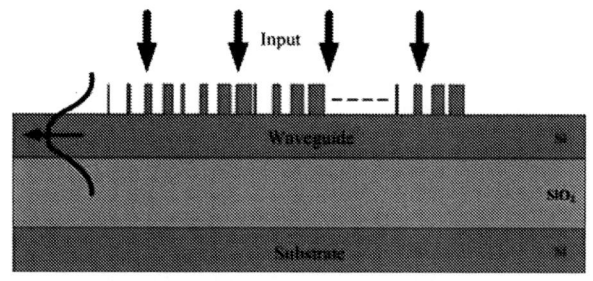

Fig. 5 Schematic of "form tapered structure" binary blazed grating coupler

A novel grating structure-"form tapered structure" (see Fig. 5), is proposed according to the theory in subwavelength area: the height and width of grating can be exchanged [8]. It decreases the transverse mode mismatch loss from grating field to waveguide by adjusting the widths of first several pillars rather than adjusting the heights of grating pillars. So the coupling efficiency can be enhanced without additional fabrication process. According to our optimization design, a vertical coupling efficiency of 59% is obtained, which is about 20% higher than conventional vertical grating coupler. The field distribution of "form tapered structure" input coupling case is

shown in figure 6. The coupling efficiency can be further improved to about 80% by multilayer reflector deposited under the grating.

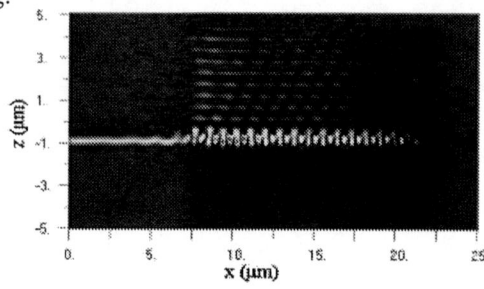

Fig.6 Field distribution of the "form tapered structure" input coupling case calculated by the software OptiFDTD 4.0. The grating length is 13.6μm and coupling efficiency is 59%.

V. MICRORING RESONATOR BIOSENSOR

With the relative ease of fabrication and potential high sensitivity, the microring resonator biosensors have been demonstrated in various material systems [9~11]. Previously, researchers attempted to obtain the high sensitivity by means of fabricating a device with high Q factors. However, our recent investigations indicate that in addition to taking efforts in fabrication, the overall sensitivity could be significantly enhanced by appropriately choosing the coupling factors, the sensing scheme, and the working wavelength.

The overall sensitivity divided into three parts is defined by equation (1):

$$S_A = S_S \cdot S_M \cdot S_W \tag{1}$$

Where S_A is the overall sensitivity which is defined as the ratio of the variation of the measured optical signal to the change of the waveguide parameter affected by the analytes [12], S_S is the sensing scheme sensitivity which determines whether monitoring the resonance shift or the output power variation is more sensitive, S_M is the mode sensitivity which represents the performance of the specific mode, S_W is the waveguide sensitivity depending on the parameters of the waveguide. These three parts are to some extent independent with each other and therefore could be studied respectively.

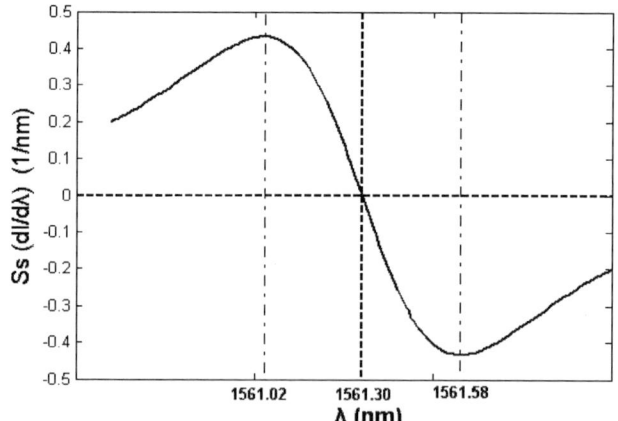

Fig.7 S_S varies as a function of working wavelength. Note that the maxima are not positioned at resonance wavelength, but symmetrically located around it.

Our research concentrates on the sensing scheme sensitivity S_S in different configurations, and two interesting phenomena are revealed:

(1) The sensitivity does not increase monotonically as the growing Q factor; (2) There exist two optimal working wavelengths at which Ss becomes maximum, and a) they are symmetrically located around the resonance, as shown by Fig 7; b) the optimal working wavelengths linearly shift towards the resonance as the sum of coupling factors increases, as shown by Fig 8.

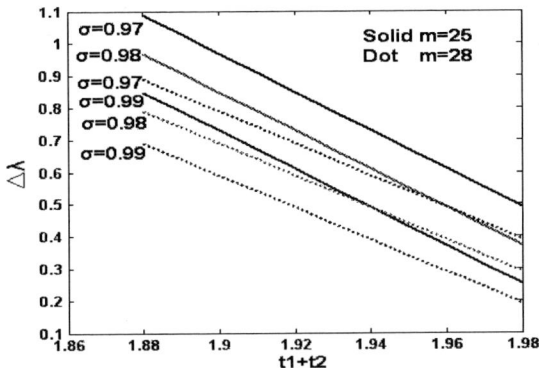

Fig.8 The shift of the optimal working wavelength ($\Delta\lambda$) varies as a function of the sum of coupling factors (t_1+t_2), mode numbers (m) and transmission factor (σ). Note that: a) the optimal working positions shift linearly towards the resonance as the sum of coupling factors increases; b) the slope of the curve reduces as the mode number increases; c) the shift of the optimal working wavelength increases as the decreasing transmission factor.

Though the explicit expression could not be obtained because the equation that controls the behaviors is transcendent, the universal phenomena described above are proved by our calculations. Therefore, aside from the high Q factor, by utilizing the rules extracted from the phenomena above, the sensing scheme sensitivity and hence the overall sensitivity could be significantly enhanced if one chooses the coupling factors, sensing scheme and working wavelength appropriately.

VI. CONCLUSIONS

Nano science and technology have been pushing sciences and technologies into ever smaller domain. They also make the silicon based monolithic integration of optoelectronic devices a reachable destination. We have taken several novel nanostructures like surface plasma, quantum well, photonic crystals, nano-grating and nonowire to design and fabricate micro/nano optoelectronic devices with high performances. Therefore, intelligent, integrated, less expensive, and highly reliable optoelectronic devices will soon march towards commodity category.

REFERENCES

[1] Y. Li, F. Qian, J. Xiang and C. M. Lieber, *"Nanowire electronic and Optoelectronic devices"*, Materialstoday, vol. 9, pp. 18–27, Oct. 2006.

[2] J. D. Joannopoulos, R. D. Meade, and J. N. Winn, *Photonic Crystals: Molding the Flow of Light*. Princeton, NJ: Princeton Univ. Press, 1995.

[3] A. Liu, R. Jones, L. Liao, D. Samara-Rubio, D. Rubin, O. Cohen, R. Nicolaescu, and M. Paniccia, *"A high-speed silicon optical modulation based on a metal–oxide–semiconductor capacitor,"* Nature, vol. 427, pp. 615–618, Feb. 2004.

[4] Y. Wang and Z. P. Zhou, *"Improved Si/Er Light Emitter by Using Two Dimensional Plasmonic Crystals,"* Proc. of SPIE, vol. 6352, pp. 63520A-1-8.

[5] J. Chen and Z. Zhou, *"Ultranarrow-band and high-quantum-efficiency photoresponse of Ge-on-Si photodetectors using cascaded-cavity structure"*, *Applied Physics Letters*, vol. 89, pp.043126, 2006.

[6] Y. Q. Jiang, W. Jiang, L. L. Gu, X. N. Chen and R. T. Chen, *"80-micron interaction length silicon photonic crystal waveguide modulator"*, *Appl. Phys. Lett.*, vol. 87, pp. 221105, Nov. 2005.

[7] Libing Zhou and Wen Liu, "Broadband polarizing beam splitter with an embedded metal-wire nanograting," *Opt. Lett.* vol. 30, pp. 1434–1436, 2005.

[8] Z. Zhou and T. J. Drabik, ''Optimized binary, phase-only, diffractive optical element with subwavelength features for 1.55 mm,'' *J. Opt. Soc. Am. A*, vol. 12, pp. 1104–1112, 1995.

[9] Ketul C. Popat, John C. Aldrige, et al, "Optical Sensing of Biomolecules Using Microring Resonators," *IEEE J. Select. Topics Quantum Electron.*, vol. 12, pp. 148-154, 2004.

[10] Junpeng Guo, Michael J. Shaw, et al, "High-Q microring resonator for biochemical Sensors," *Proc. Of SPIE*, vol. 5728, pp. 83-92, 2005.

[11] C. Y. Chao, L. J. Guo, "Biochemical Sensors Based on Polymer Microrings with Sharp Asymmetrical Resonance," *Appl. Phys. Lett.*, vol. 83, pp. 1527-1529, 2003.

[12] C. Y. Chao, L. Jay Guo, "Design and Optimization of Microring Resonators in Biochemical Sensing Applications," *Journal of ligthwave technology*, vol. 24, pp. 1395-1402, 2006.

In-situ Opening Aligned Carbon Nanotubes and Applications for Device Assembly and Field Emission

Lingbo Zhu[1,2], Dennis W. Hess[1], C. P. Wong[2]
[1]School of Chemical & Biomolecular Engineering
[2]School of Material Science & Engineering
Georgia Institute of Technology
771 Ferst Drive NW
Atlanta, GA 30332

Abstract—**Carbon nanotubes (CNTs) have been proposed for applications in microelectronic applications, especially for electrical interconnects and nanodevices, due to their excellent electrical, thermal and mechanical properties. Usually, the CNTs produced by arcing, laser ablation or chemical vapor deposition (CVD) are inevitably close-ended. Due to the weak coupling of the individual walls and close ends, it leads to conclusions that only the outer wall of multi-walled CNT is contributed to the current-carrying capacity. However, recent research shows that each wall of the multi-walled CNTs contributes to the saturation current to obtain a very high current-carrying capacity, i.e., the multichannel electron transport could be achieved by opening multi-walled CNTs. The previous process to open the CNTs can't be applied to the aligned CNTs, since they will damage the original alignment of CNTs. In this paper, we for the first time report a simple process to achieve simultaneous CNT growth and opening of the CNT ends, while keeping alignment of the original CNT films/arrays. The addition of relatively low reactivity oxidizing agents (water) into the reaction furnace has been demonstrated the feasibility. The as-grown CNTs were characterized by high resolution transmission electron microscopy (HRTEM), scanning electron microscopy (SEM). Also, we proposed using novel CNT transfer technology, enabled by open-ended CNTs, to circumvent the high carbon nanotube (CNT) growth temperature and poor adhesion with the substrates that currently plague CNT implementation. The process is featured with separation of high-temperature CNT growth and low-temperature CNT device assembly. Field emission testing of the as-assembled CNT devices is in a good agreement with the Fowler-Nordheim (FN) equation, with a field enhancement factor of 4540. This novel technique shows promising applications for positioning CNTs on temperature-sensitive substrates, and for the fabrication of field emitters, electrical interconnects, thermal management structures in microelectronics packaging**

I. INTRODUCTION

Carbon nanotubes (CNTs) have attracted great interest due to their extraordinary structural, electrical, and mechanical properties, and their wide range of potential applications [1]. The CNTs can be either metallic or semiconducting, depending upon how the graphite layer is wrapped into a cylinder [2, 3]. Many reports have been published on the utilization of the individual or assembly of the nanotubes. For instance, the nanotubes were used for advanced scanning probe [4], room temperature transistor [5], and field emitter arrays for flat panel display [6].

For applications of the nanotubes in microelectronics, the most interesting features are the ballistic transport of electrons and the extremely high thermal conductivity along the tube axis [7]. Metallic CNTs show ballistic conductivity at room temperature [8]. Electrons transport ballistically (without scattering) in metallic SWNTs and MWNTs over reasonable lengths (~1µm), thereby enabling CNTs to carry very high currents (> $10^9 A/cm^2$) without electromigration failure [9]. Phonons also propagate easily along the nanotubes [9]. The measured thermal conductivity of an individual MWNT at room temperature is >3000 W/m·K [10], which exceeds the conductivity of diamond (2000 W/m·K). Based on these advantageous properties of CNTs, researchers have reported the integration of CNTs into electrical interconnect applications [11-13].

However, the resistance of a single ballistic SWNT less than 1 µm long is about 6.5 kΩ with perfect contacts [14], while ballistic transport in MWNTs with a resistance of 12.9 kΩ has also been reported [15]. The high resistance of an individual CNT indicates that an array of thousands of parallel CNTs will be necessary for interconnect applications. Previous studies have demonstrated the synthesis of aligned nanotubes. Ren et al. demonstrated that large areas of aligned carbon nanotubes could be grown on glass substrates using plasma-enhanced hot filament chemical vapor deposition [16]. However, the CNTs' area distribution density is not high enough to decrease the electrical resistance. In addition, a method for the growth of CNT arrays by the gas-phase delivery of a xylene-ferrocene mixture has been demonstrated [17, 18]. In this process, Fe catalyst particles are produced continuously during CNT growth. These conductive particles limit the applications of CNTs in microelectronic device fabrication because they can cause electrical shorts in the circuits or between interconnects. Therefore, for electrical interconnect applications, the growth of pillars of thousands of parallel CNTs in very high density and high purity is still necessary.

The electronic properties of perfect multi-walled carbon

1-4244-0816-4/06/$25.00 ©2006 IEEE

nanotubes (MWCNTs) are similar to those of single-walled carbon nanotubes (SWCNTs) [19]. For close-ended CNTs, the coupling between the graphite cylinders is weak (layer distance is 0.34 nm), so only the outer shell of MWCNTs contribute to the current transport. However, recent studies have demonstrated that the internal walls of MWCNTs can participate in electrical transport, thereby enabling large current-carrying capacity [20]. It implies that multichannel ballistic transport could be achieved if the caps of the CNTs are removed; CNT electrical conductance should therefore be dramatically improved. Such achievements may then allow CNTs to serve as conductive nanowires and thus replace copper and aluminum films used in state-of-the-art circuits; such nanowires are less susceptible to electromigration under high current density than are Cu and Al.

CNTs produced by arcing, laser ablation or chemical vapor deposition (CVD) are inevitably close-ended. The nanotubes can be opened by post-treatment process, such as using oxygen [21] and carbon dioxide [22], to etch away CNT caps; however, the nanotube walls are inevitably damaged. As a result, CNT electrical and mechanical properties are degraded. Realization of high-quality open-ended CNT synthesis requires either a novel post-treatment process to open CNTs, or an alteration of the CVD process to allow in-situ high-quality open-ended CNT growth without the need for subsequent processes. In-situ growth of open-ended CNTs is desirable, since it is cost-effective. Our intent is to develop a novel process to open the nanotubes in-situ in order to study the corresponding CNT properties while maintaining CNT film alignment. In this paper, we report a novel process for in-situ opening CNTs by water-assisted selective etching. By taking advantage of such aligned open-ended CNT structure, we build CNT architecture using novel CNT transfer technology. The success of this methodology is reflected in the performance of the assembled CNT field emitters. This process may offer a new paradigm for transferring and integrating CNTs onto integrated circuits (ICs) as well as other moduli in microelectronic packaging systems, since the approaches used circumvent the high CNT growth temperature and poor adhesion that currently plague CNT implementation.

II. EXPERIMENTAL

Ppm levels of water can etch the ends of the nanotubes as described previously [23]; only a brief description of the process will be presented here. The substrates used in this study were (001) silicon wafers coated with SiO_2 (500nm) by thermal oxidation. The catalyst layers of Al_2O_3 /Fe were formed on the silicon wafer by sequential e-beam evaporation. CVD growth of CNTs was carried out at 775°C with ethylene as the carbon source, and hydrogen and argon as carrier gases. The water vapor concentration in the CVD chamber was controlled by bubbling a small amount of argon gas through water held at 22 °C. Ethylene flowed into the CVD system for a preset time, after which the flow was terminated; it followed by 5 min of only water, argon and hydrogen flow, which was used to selectively etch the nanotube tips and carbon atoms at the interfaces

between the nanotubes and catalyst particles.

Samples for transmission electron microscopy (TEM) were prepared by sonicating a trace amount of synthesized carbon nanotubes in methanol. A few drops of the suspension were placed onto a TEM grid. TEM (JEOL 400EX) analysis was performed at 400 kV. Scanning electron microscopy (SEM) characterization was carried out on a JEOL 1530 equipped with a thermally assisted field emission gun operating at 10 keV. Spatially resolved energy-dispersive X-ray spectroscopy (EDS) was used to examine the purity of carbon nanotubes, as well as the catalyst layer at specific locations along the length of the nanotubes.

III. RESULTS AND DISCUSSIONS

A. Close-ended CNT film and array growth

For the growth of close-ended CNTs, no water is introduced into CVD chamber during CNT growth. When the growth temperature exceeded 650°C, deposition of a black film was observed on the substrate after the samples were removed from the furnace. Typical SEM images of these samples obtained in a cross-section view are shown in Figure 1. Figure 1a shows an SEM image of aligned nanotubes grown normal to the substrate at 775 °C for 10 min. Figures 1b shows the nanotubes synthesized at 800°C for 10 min.

(a)
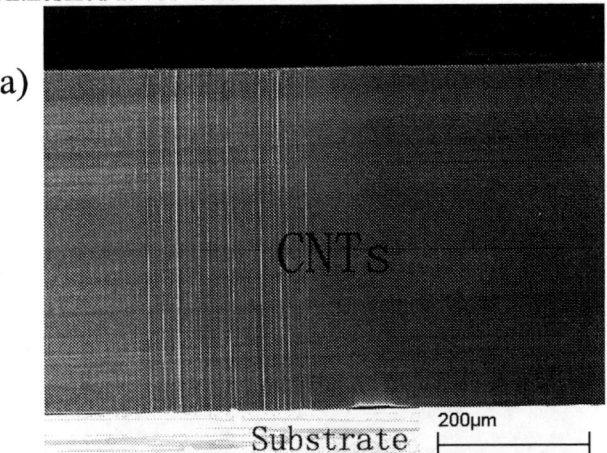

(b)
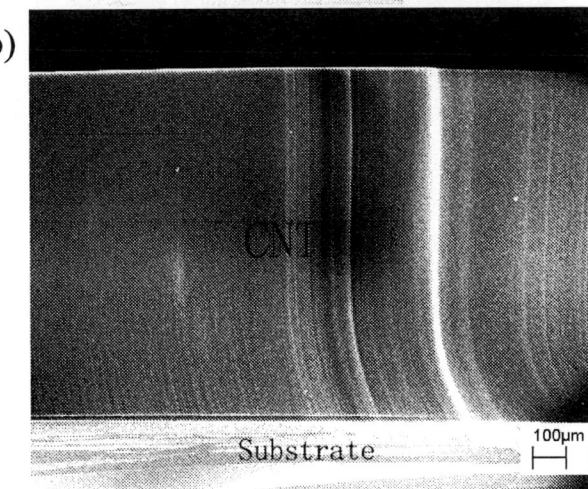

Figure 1. SEM images of carbon nanotube films grown at temperatures of (a) 750, and (b) 800°C for 10 min.

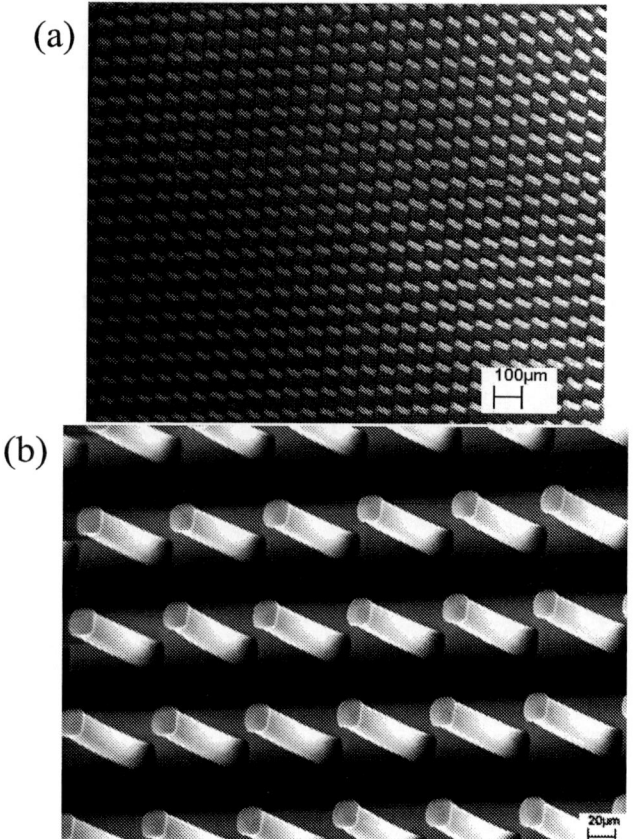

(a)

A higher magnification SEM image of the sample is shown in Figure 2a. This image demonstrates that no iron or carbon particles protrude from the nanotubes; also, it can be seen clearly that the film consists of well-aligned CNTs with high area distribution density. Measurement of the CNT average diameter (10 nm) and pitch size (25 nm) using an SEM allowed an estimate of the nanotube density (>2000 μm^{-2}). Figure 2b shows a TEM image of the CNT bundles. CNT diameters range from 8 to 13 nm, with an average diameter of 10 nm.

(b)

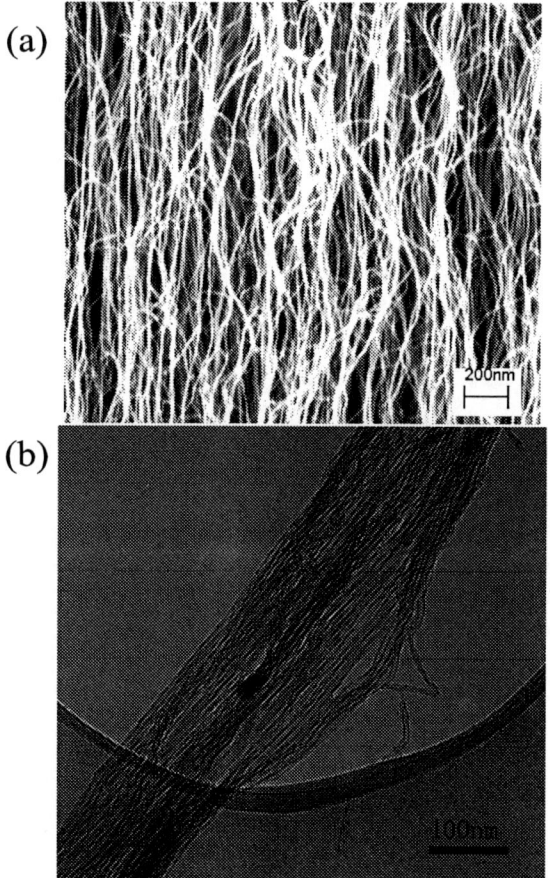

Figure 3. SEM images of (a) as-grown well-aligned CNT pillars on 20 μm catalyst islands; (b) magnified image of (a).

B. Open-ended CNT growth

We reported the growth of CNT stacks up to 10 layers by water-assisted selective etching [23]. Figure 5 shows that five-layered CNT stacks could be obtained by repeating the growth-etching cycle five times. Growth conditions for the CNTs shown in Figure 5a are 80 sec of ethylene flow followed by 5 min with the ethylene flow terminated in every cycle at 775 °C with ethylene, argon and hydrogen flow rates of 150, 350, and 180 sccm, respectively; the Si/SiO₂ substrate had 3 nm Fe layer deposited onto a 15 nm Al₂O₃ layer. The CNT films in Figure 2a were partially peeled off using tweezers to demonstrate the layered structures of CNT films, as shown in Figure 2b

Figure 2. (a) High magnification SEM image of nanotube film; (b) TEM image of bundled CNTs.

By using a lift-off process, a patterned layer of Fe catalyst can be formed on the substrate for the CVD growth of aligned CNT arrays. Aligned CNTs grow readily from catalyst patterns into well-defined vertical structures, as shown in Figure 3. The CNT growth conditions for Figure 3a are 700°C for 3 min. The images demonstrate that nearly vertically-aligned CNT arrays can be formed on the substrate without entanglement with neighboring bundles. Figure 3b is magnified SEM image of Figure 3a. The successful aligned nanotube array growth and easy length control make this process very promising for electric interconnect applications. The observed height and aspect ratios of CNT pillars by this process should satisfy most interconnect applications.

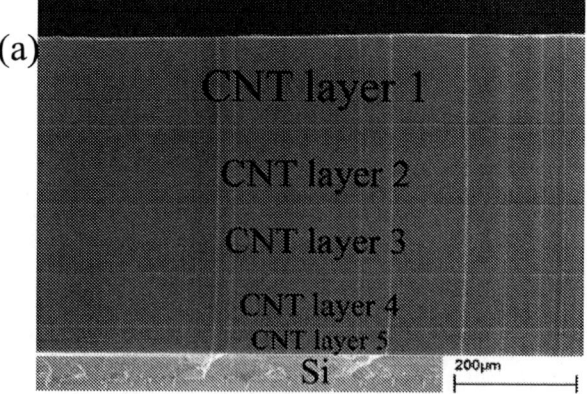

(a)

14

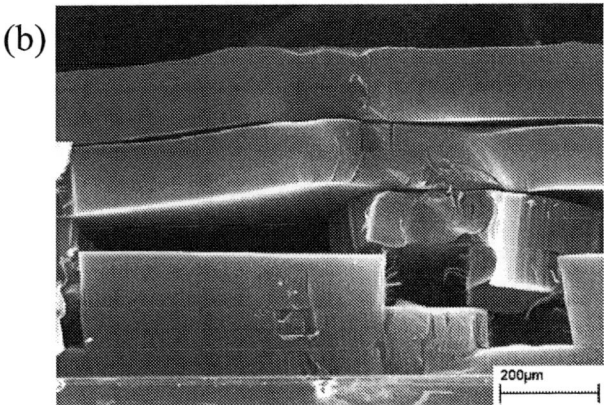

Figure 4. (a) Cross-sectional SEM images of 5-layered CNT films; (b) scratched CNT films in (a) to show the layered structures of CNT films.

We believe that the relatively small amount of water etches the ends of the nanotubes because more defect structures and thus high reactivity exist at the ends of the nanotubes, though it may be possible that the water also attacks the defects along the nanotube walls. HRTEM image shows CNTs with one open end as indicated in Figure 4a; Figure 4b shows the opposite (open) end of the same nanotube as in Figure 4a. We have examined CNTs from numerous growth runs, with dozens of HRTEM images. Each image has shown open ended structures of the as-grown CNTs.

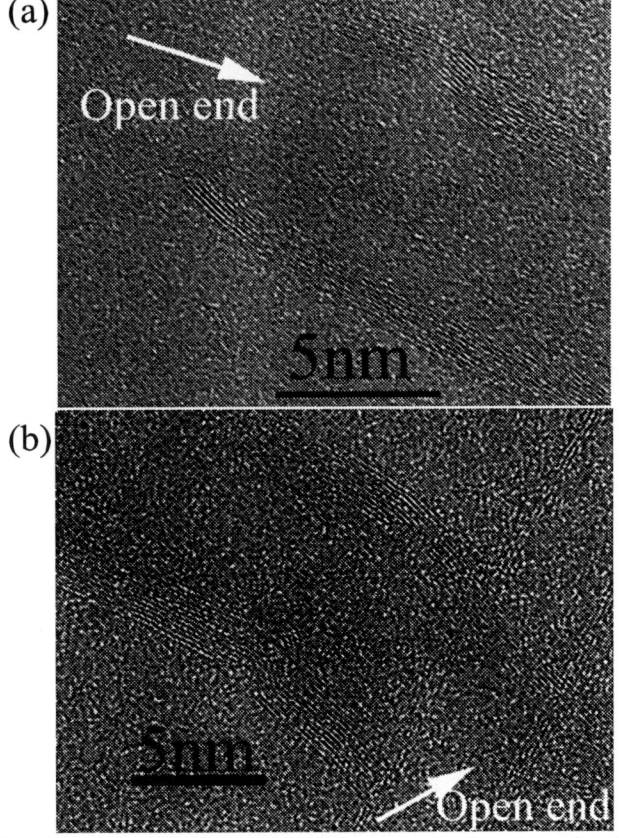

Figure 5. (a) and (b) HRTEM images of the two ends of one nanotube, showing that the two ends are open.

C. CNT transfer technology

For electronic device applications, chemical vapor deposition (CVD) methods are particularly attractive due to characteristic CNT growth features such as selective spatial growth, large area deposition capability and aligned CNT growth. However, the CVD technique suffers from several drawbacks. One of the main challenges for applying CNTs to circuitry is the high growth temperature (>600°C). Such temperatures are incompatible with microelectronic processes, which are typically, performed below 400-500 °C in back-end-of-line fabrication sequences. Another issue is the poor adhesion between CNTs and the substrates, which will result in long term reliability issues and high contact resistance. To fabricate microelectronics devices that incorporate CNT blocks, the CNTs should be selectively positioned and interconnected to other materials such as metal electrodes or bonding pads. To overcome these disadvantages, we propose a methodology that we term "CNT transfer technology", which is enabled by open-ended CNT structures. This technique is similar to flip-chip technology as illustrated schematically in Figure 6 [24].

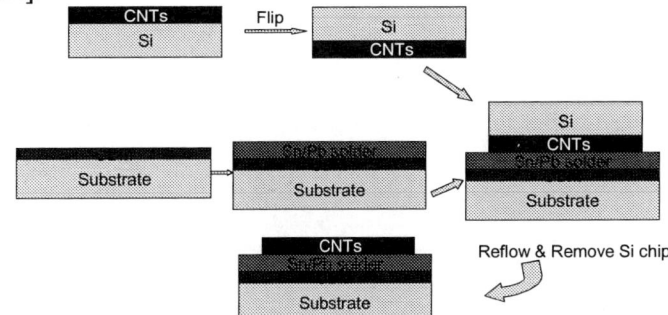

Figure 6. Schematic diagram of "CNT transfer technology". UBM: under bump metallization. See the text for detailed explanations.

The substrates can be FR-4 boards coated with copper foil or other materials and moduli, such as heat sinks. To improve the adhesion and wetting of solder on the substrates, the under bump metallization (UBM) layers are sputtered onto the substrate metallization. The eutectic tin-lead paste is then stencil-printed on the UBM. The silicon substrates with CNTs are flipped and aligned to the corresponding copper substrates, and reflowed in a seven-zone BTU reflow oven at higher temperatures (peak temperature at 270 °C) than those typically used (220 °C) to simultaneously form electrical and mechanical connections. This process is straightforward to implement and offers a strategy for both assembling CNT devices and scaling up a variety of devices fabricated using nanotubes (e.g., flat panel displays). This process could overcome the serious obstacles of integration of CNTs into integrated circuits and microelectronic device packages by offering low process temperatures and improved adhesion of CNTs to the substrates. Figure 7a indicates that the entire CNT film (1.5 cm × 1.5 cm) is transferred to the substrate (2.54 cm × 2.54 cm), since no trace amount of CNTs are evident on the silicon chip. This result is in stark contrast to the same process wherein the CNTs are closed-ended, in that only a fraction of the CNT film is transferred to the substrate (shown in Figure 7b), indicating that the adhesion of the attached nanotube films is poor. In addition, upon drop testing, the closed-ended CNT films detached from the substrate, while open-ended CNTs did not. Such observations indicate that the wetting properties of solders on the aligned open-ended CNT film have been improved. We believe that the open channels of nanotubes assist the adhesion between nanotubes and solders due to strong capillary forces that draw solder inside the CNTs.

Figure 7. (a) Photograph of open-ended CNT film transferred onto the copper substrate coated with eutectic tin-lead solder. (b) Photograph of closed-ended CNT film which has only partially transferred onto the copper substrate coated with eutectic tin-lead solder.

Figure 8a is a cross-sectional SEM image of well-aligned open-ended CNT films transferred onto the copper substrate. Any detachment of CNT film from the substrate is not found. To qualitatively demonstrate the bonding strength of the CNTs on the copper substrate that results from the solder reflow process, a section of the assembled CNTs was pulled from the surface by tweezers. Figure 8b shows the demarcation between the broken CNTs and the intact and connected ones. When pulled from the substrate, the CNTs break along the axis rather than at the CNT-solder interface. The excellent mechanical bonding strength of CNTs on the substrate anchor the CNTs and thereby makes decreasing electrical/thermal contact resistance possible.

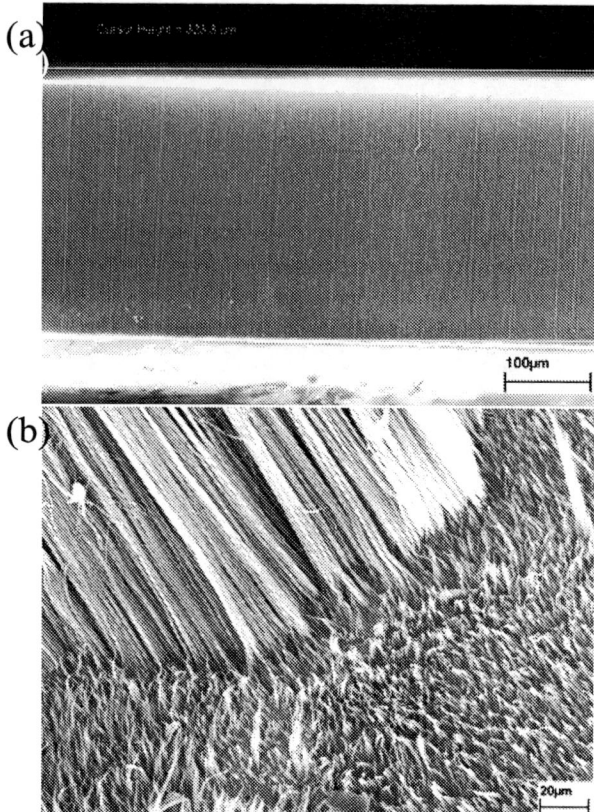

Figure 8. (a) A cross-sectional SEM image of well-aligned open-ended CNT films transferred onto the copper substrate. The eutectic tin-lead solder is used to interconnect CNTs and the copper substrate. (b) SEM of the copper substrates on which the CNTs were assembled after some CNTs were pulled from the surface by tweezers; this shows the excellent mechanical bond strength of CNTs transferred to the copper substrate by the solder reflow process.

To explore the electrical properties of CNTs connected by solders on the copper substrates, field emission characterization of the as-prepared assembly has been performed. The height of the nanotube films is ~323 µm with diameters in the range of 10 to 20 nm. We measured the (cathodic) electron emission from 1.5 cm by 1.5 cm well-aligned open-ended CNT films shown in Figure 7a at room temperature and in a vacuum chamber below 10^{-5} Torr. The spacing between the CNT tip and the anode (phosphor-coated ITO glass) was ~180 µm and was maintained by a poly (tetafluoroethylene) (PTFE) spacer. The measured current density (mA/cm²) as a function of electric field (V/µm) is shown in Figure 9. A typical turn-on field, which produces a current density of 10 µA/cm², is ~1.8V/µm, while the emission current density of 1 mA/cm² requires an applied field of ~2.74 V/µm. The small turn-on field is consistent with literature data of 1.5-2 V/µm observed in CVD-grown dense CNT films [25]. At an electric field of 3.4 V/µm, the assembled CNT field emitters emit a current density of 5 mA/cm², which is a remarkably large value considering the distance between cathode and anode. A plot of $\ln(I/V^2)$ versus $1/V$ yields a straight line in a good agreement with the Fowler-Nordheim (FN) expression; this agreement demonstrates that the current originates from field emission (i.e. field emission process) [26]. Furthermore, the quality of fit to the Fowler-Nordheim

16

expression implies good nanotube/substrate electrical contact. The slope of the FN plot can be used to calculate the field enhancement factor β. The Fowler-Nordheim equation can be written as [27]:

$$I = A \frac{1.42 \times 10^{-6}}{\phi} \beta^2 (\frac{V}{d})^2 \exp(\frac{10.4}{\sqrt{\phi}}) \exp(-\frac{B\phi^{1.5}d}{\beta V})$$

where I is the emission current (A), A the emission areas (m^2), V applied voltage, d the distance between CNT tips and anode (m), ϕ the work function (eV), and B constant (6.44×10^9, VeV$^{-1.5}$m^{-1}). When ln(I/V^2) is plotted versus 1/V, the slope of this linear formulation is given by $-B\phi^{1.5}d / \beta$. Assuming that the work function is 5.0 eV [28], the derived field enhancement factor is calculated to be 4540, which is sufficient for application in field emission displays.

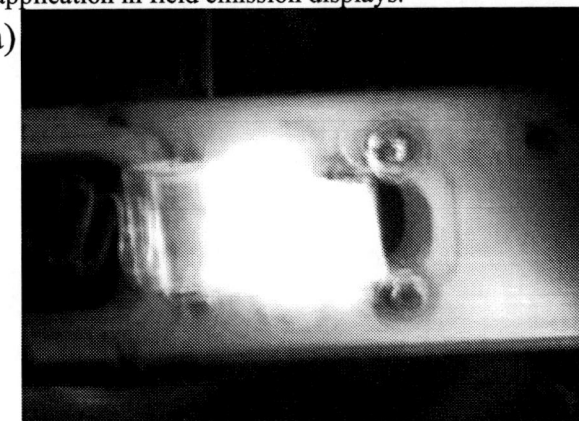

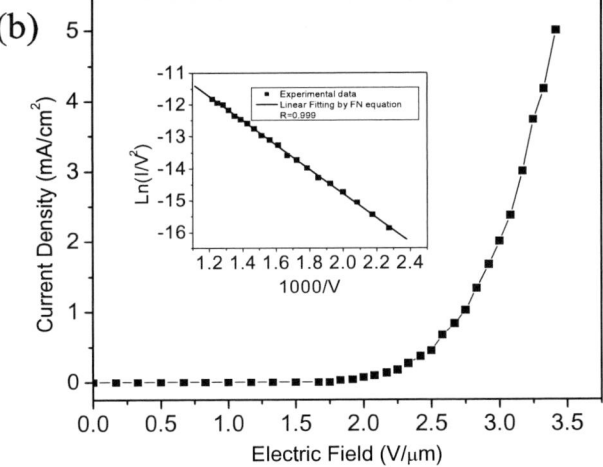

Figure 9. (a) Emission pattern of the as-assembled CNTs by applying electrical field of 3.0 V/μm. (b) Field emission measurements of CNT films in (a) at room temperature. The inset shows a Fowler-Nordheim plot, which indicates that the transferred CNTs demonstrate good field emission characteristics.

IV. CONCLUSIONS

We have demonstrated successfully a simple but efficient method for the growth of well-aligned CNT films and pillars with high density under a wide range of growth parameters. At 800°C, the nanotube films grow at a high average rate of 100 μm/min, and their length can achieve 1 mm in 10 min. Also, we reported an in-situ method to manufacture well-aligned open-ended CNTs. The open-ended structures are the key to the successful assembly of CNTs on substrates by a solder reflow process. This process is compatible with current microelectronics fabrication sequences and technology. The distinctive CNT-transfer-technology features are separation of high-temperature CNT growth and low-temperature CNT device assembly. Overall, the advantages of CNT transfer technology are embodied in the low process temperature, adhesion improvement and the feasibility of transferring CNTs to different substrates. Field emission testing of the as-assembled CNT devices indicates good field emission characteristics, with a field enhancement factor of 4540. The testing results are in agreement with the Fowler-Nordheim expression, which implies good electrical contact between CNTs and solder, and a very small voltage drop across CNT/solder interfaces. CNT transfer technology shows promising applications for positioning of CNTs on temperature-sensitive substrates, and for the fabrication of field emitters, electrical interconnects, thermal management structures in microelectronics packaging

ACKNOWLEDGMENTS

We would like to thank NSF for funding support (DMI-0422553). We also thank Dr. Yong Ding for HRTEM examinations.

REFERENCES

[1] Dai, H.J., Kong J., *et al.* "Controlled Chemical Routes to Nanotube Architectures, Physics, and Devices", *J. Phys. Chem. B*, Vol. 103, No. 51 (1999), pp. 11246-11255.

[2] Hamada, N., Sawada, S.-I., Oshiyama, A., "New One-Dimensional Conductors: Graphite Microtubule", *Phys. Rev. Lett.*, Vol. 68, No. 10 (1992), pp. 1579-1581.

[3] Saito, R., Fujita, M., Dresselhaus, G., Dresselhaus, M. S., "Electronic Structure of Chiral Graphene Tubules", *Appl. Phys. Lett.*, Vol. 60, No. 18 (1992), pp. 2204-2206.

[4] Hafner, J. H.; Cheung C. L.; "High-yield Assembly of Individual Single-Walled Carbon Nanotube Tips for Scanning Probe Microscopy", *J. Phys. Chem. B*, Vol. 105, No. 4 (2001), pp. 743-746.

[5] Tans, S. J.; Verschueren, A. R. M.; Dekker, C. "Room-Temperature Transistor Based on a Single Carbon Nanotube", *Nature*, Vol. 393, (1998), pp. 49-51.

[6] Bonard, J. M., Salvetat, J. P., Stöckli, T., Forro, L., Chatelain, A., "Field Emission From Carbon Nanotubes: Perspective for Applications and Clues to the Emission Mechanism", *Appl. Phys. A*, Vol. 69, No. 3 (1999), pp.245-254.

[7] Hoenlein, W.; Kreupl, F; "Carbon Nanotube Applications in Microelectronics", *IEEE T. Compon. Pack. T.*, Vol. 27, No. 4 (2004), pp. 629-634.

[8] S. Frank, P. Poncharal, Z. L. Wang, and W. A. de Heer, "Carbon Nanotube Quantum Resistors", *Science*, Vol. 280, No. 5370 (1998), pp. 1744-1746.

[9] Baughman, R H.; Zakhidov, A. A, *et al*, "Carbon Nanotubes – the Route Toward Applications", *Science*, Vol. 297, No. 5582 (2002), pp. 787-792.

[10] Kim, P., Shi, L., McEuen, P. L., "Thermal Transport Measurements of Individual Multiwalled Nanotubes", *Phys. Rev. Lett.*, Vol. 87, No. 21 (2001), pp. 215502-1-4.

[11] Graham, A. P., Duesberg, G. S. *et al*, "Towards the Integration of Carbon Nanotubes in Microelectronics", *Diam. Relat. Mater.*, Vol. 13, No. 4-8 (2004), pp. 1296-1300.

[12] Kreupl, F.; Graham, A. P. *et al*, "Carbon Nanotubes in Interconnect Applications", *Microelectron. Eng.*, Vol. 64, No. 1-4 (2002), pp. 399-408.

[13] Homma, Y., Yamashita, T., Kobayashi, Y., *et al*, "Interconnection of Nanostructures Using Carbon Nanotubes", *Physica B*, Vol. 323, No. 1 (2002), pp. 122-123.

[14] McEuen, P. L., Fuhrer, M. S., Park, H., "Single-Walled Carbon Nanotube Electronics", *IEEE T. Nanotechnol.*, Vol. 1, No. 1 (2002), pp. 78-85.

[15] Dresselhaus, M. S., *et al.*, Carbon nanotubes: Synthesis, Structure, Properties and Applications, Springer (Berlin, 2001), pp. 147-168.

[16] Ren, Z. F., *et al*, "Synthesis of Large Arrays fo Well-Aligned Carbon Nanotubes on Glass", *Science*, Vol. 282, No. 5391, pp.1105-1107.

[17] Wei, B. Q., *et al*, "Assembly of Highly Organized Carbon Nanotube Architectures by Chemical Vapor Deposition", *Chem. Mater.*, Vol. 15, No. 8 (2003), pp. 1598-1606.

[18] Cao, A.; Baskaran, R.; Frederick, M. J.; Turner, K.; Ajayan, P. M.; Ramanath, G., "Direction-selective and length-tunable in-plane growth of carbon nanotubes", *Adv. Mater.*, Vo. 15, No. 13 (2003), pp. 1105-1109.

[19] Collins, P. G., Arnold, M. S., Avouris, P., "Engineering Carbon Nanotubes and Nanotube Circuits Using Electrical Breackdown", *Science*, Vol. 292, No. 5517 (2001), pp. 706-7096.

[20] Li, H. J., Lu, W. G., Li, J. J., Bai, X. D., Gu, C. Z., "Multichannel Ballistic Transport in Multiwall Carbon Nanotubes", *Phys. Rev. Lett.*, Vol. 95, No. 8 (2005), pp. 086601-1-4.

[21] Ajayan, P. M., *et al,* "Opening Carbon Nanotubes With Oxygen and Implications for Filling", *Nature*, Vol. 362 (1993), pp. 522-525.

[22] Tsang, S. C., et al, "Thinning and Opening of Carbon Nanotubes by Oxidation Using Carbon Dioxide", *Nature*, Vol. 362 (1993), pp. 520-522.

[23] Zhu, L., Xiu, Y., Hess, D. W., Wong, C. P., "Aligned Carbon Nanotube Stacks by Water-Assisted Selective Etching", *Nano Lett.*, Vol. 5, No. 12 (2005), pp. 2641-2645.

[24] Wong, C. P. *et al*, US Patent (pending).

[25] Yu, W. J., *et al*, "Patterned Carbon Nanotube Field Emitter Using the Regular Array of An Anodic Aluminium Oxide Template", *Nanotechnol.*, Vol. 16 (2005), pp. S291-S295.

[26] Sohn, J. I., *et al*, "Patterned selective growth of carbon nanotubes and large field emission from vertically well-aligned carbon nanotube field emitter arrays", *Appl. Phys. Lett.*, Vol. 78, No. 7 (2001), pp. 901-903.

[27] Bonard, J. M., *et al*, "Degradation and Failure of Carbon Nanotube Field Emitters", *Phys. Rev. B*, Vol. 67 (2003), pp. 115406-115415.

[28] Lee, O. J.; Lee, K. H, "Fabrication of Flexible Field Emitter Arrays of Carbon Nanotubes Using Self-Assembly Monolayers", *Appl. Phys. Lett.*, Vol. 82, No. 21 (2003), pp. 3770-3772.

Radio-Frequency and Millimeter-Wave Photonic Techniques for Broadband Communications and Sensor Networks

Jianping Yao, *Senior Member, IEEE*, Xiupu Zhang, and Raman Kashyap, Ke Wu, *Fellow, IEEE*

Abstract—Recent advances in radio-frequency (RF) and millimeter-wave (mmW) photonics are reviewed with emphasis on our research in broadband radio-over-fiber system architectures, photonic generation and processing of RF and mmW signals, and optoelectronic device design for mmW photonics applications. Future system design challenges are also discussed. In particular, the concept of substrate integrated circuits (SICs) is shown to have great potentials in the design of highly integrated mmW circuits as well as traveling-wave optoelectronic devices, which are critical to the development of future RF and mmW photonic systems for broadband wireless access and sensor applications.

I. INTRODUCTION

Wireless communication systems operating in the millimeter-wave (mmW) bands of 30 and 60 GHz have recently received considerable interest as they offer a high potential for broadband and secure data transmission. A microcellular or picocellular system architecture of 2-200 meters in which mmW carriers that deliver high bit rate signals over optical fiber or radio-over-fiber (RoF) to remote wireless base stations or vice-versa will be the most attractive economical and universal solution for seamless broadband multimedia communications and wireless sensor networks. The fundamental purpose of current research in this direction is to investigate cost-efficient RoF network architectures, maintenance-free remote RoF base-stations, and low-cost integrated mmW, optical and optoelectronic devices for future RoF networks. A maintenance-free base-station means a base station that does not require local electrical and optical sources, which is powered by a central or remote station.

In this paper, we will review a number of cost-effective techniques developed recently that enable optical generation, distribution, and processing of mmW signals for broadband wireless access networks and sensor networks applications. The paper consists of 6 sections. In Section 2, novel modulation and demodulation schemes based on DWDM techniques to efficiently distribute mmW signals over optical fiber will be discussed. In Section 3, techniques to generate mmW signals for base-station applications will be presented, which include phase locking of two ultra-quiet lasers and mmW source generation using external modulation techniques. The techniques to generate a mmW source using a dual-wavelength fiber laser, and a rational harmonic mode locked fiber laser are presented. The implementation of mmW frequency tripling based on four wave mixing (FWM) in a semiconductor optical laser (SOA) is

also discussed in this section. In Section 4, we will show some new all-optical microwave signal processing (all-optical microwave mixing and filtering) techniques which would find applications in RoF networks. Low-cost substrate integrated circuits (SICs) that potentially allow full and single-substrate system integration will be presented in Section 5. Finally, a conclusion in Section 6 is drawn with a discussion on future system design challenges.

II. MODULATION AND DEMODULATION SCHEMES BASED ON DWDM TECHNIQUE

In a RoF system, optical subcarrier modulation is generally used to multiplex multiple channels of mmW signals, which are modulated on an optical carrier, distributed over optical fiber for broadband wireless access. In this section, we will present our techniques based on optical subcarrier modulation (SCM), optical multiplexing and demultiplexing of mmW-band subcarrier-modulated optical signals, and optical mmW generation and signal reception at a base station. The nonlinear distortions in a RoF system and techniques to suppress them will also be discussed.

A. Optical Subcarrier Modulation

Normally, a microwave carrier is imposed on an optical carrier and the microwave carrier is modulated with a digital electrical signal. Thus there are two carriers, the optical and the microwave carrier. The microwave carrier is referred to as the subcarrier in the optical domain. Transmission of an optical carrier with its subcarrier over fiber is referred to as RoF with SCM. Practically, it is straightforward to obtain an optical SCM by directly driving a laser diode or an external modulator by a microwave signal. For optical intensity modulation, two optical subcarriers are generated, separated from the optical carrier by a frequency of the microwave carrier, ω_{RF}. Because of two subcarriers, it is referred to as double sideband (DSB) SCM, with one upper sideband (USB) and one lower sideband (LSB). Since the bandwidth of a laser diode is limited to 10 GHz, for mmW applications an external modulator is usually used, such as a LiNbO3 Mach Zehnder modulator (MZM) or an optical electro-absorption modulator (EAM).

Thus, DSB in SCM suffers from fiber chromatic dispersion induced power fading since the two subcarriers, separated by two times of the microwave frequency, have a large walk-off after fiber transmission. When the phase difference of the two subcarriers after fiber transmission is π, direct photodetection of the optical signal generates a null since the beating between the LSB with the optical carrier completely cancels the beating between the USB with the optical carrier. To reduce fiber chromatic-dispersion induced power penalty, a simple way is to use a single sideband (SSB) SCM, eliminating the cancellation at the photodetector. A technique to realize SSB SCM is to use an optical notch filter to remove one of the two subcarriers after optical modulation. The use of a cost-effective narrow band fiber Bragg grating (FBG) to realize SSB SCM has been recently demonstrated [1]. Another important technique for SSB SCM generation is to use a two-port external modulator [2], in which a RF signal is split into two branches, which drive the two arms of the MZM with $\theta = \pm \pi/2$ phase delay in one branch. When $\theta = +\pi/2$ ($-\pi/2$) is used, only the LSB (USB) subcarrier will be generated.

To improve the optical spectral efficiency, an optical carrier should carry multiple optical subcarriers with each conveying its own information. As an example, we consider an optical carrier for transporting two subcarriers. When using SSB SCM, two RF signals at Ω_1 and Ω_2 are first combined electrically and then split into two branches by a hybrid splitter, to drive the two arms of the MZM; one branch has a phase shift of $\theta = \pm \pi/2$. At the output of the MZM, two subcarriers are located either in the LSB or the LSB [2]. Alternatively, the two subcarriers can be located in tandem of the optical carrier, i.e. tandem SSB (TSSB) SCM. To obtain TSSB SCM, two RF signals, at Ω_1 and Ω_2, drive a four port 90^0 hybrid splitter. The two hybrid outputs drive a dual-arm MZM, biased at the quadrature point [3]. At the output of the MZM, the two subcarriers at Ω_1 and Ω_2 are located at the USB and the LSB.

B. Optical Multiplexing and Demultiplexing

DWDM technique can be applied to mmW-band RoF systems to increase the number of optical carriers transmitting over a single fiber. When DWDM channel spacing is 100 GHz or higher, the optical carrier and its subcarriers can be within the passband of optical multiplexers and demultiplexers. Thus a commercially available multiplexers and demultiplexers can be applied to mmW RoF systems. However, when the channel spacing is 50 GHz or smaller, a commercially available multiplexers and demultiplexers cannot be used directly, since the frequency interval between the optical carrier and its subcarriers is greater than the channel spacing. A straightforward method to overcome the problem is to modify the optical multiplexers and demultiplexers [4], which unfortunately significantly increases the cost of the RoF systems.

Recently, we proposed a novel technique to solve this problem by using conventional DWDM multiplexers and demultiplexers without using optical interleaving [5]. In the proposed approach, two lasers with a small wavelength difference, phase locked and polarization aligned, are located at a central station (CS) to connect the CS with each base station

(BS), one laser for transmitting and the other for detection (the remote local oscillator). For a conceptual illustration, we consider a DWDM RoF system with channel spacing of 12.5 GHz and a subcarrier frequency in the 30-GHz mmW-wave band. In the downlink, a SSB subcarrier is used with a low microwave frequency imposed on an optical carrier at the CS, and a mmW-band signal is obtained at each BS using direct photodetection by the SSB subcarrier which beats with the remote oscillator signal. In the uplink, the received mmW-band signal at each BS is imposed on the two optical carriers simultaneously, one optical carrier with the closest SSB subcarrier is optically filtered out and fed into in the uplink transmission fiber without frequency interleaving; the electrical signal with a low intermediate frequency (IF) can be photodetected directly at the CS. Such a RoF system has simple, cost-effective, and maintenance reduced BSs, and is immune to laser phase noise in principle.

C. Millimeter-Wave Generation and Reception at Base Stations

We proposed a novel technique with remote local optical oscillators which are located in the CS for the generation of mmW frequencies at BSs. At BSs, the beating of the remote local optical oscillator signal and the subcarrier will directly generate a signal at the mmW band. The mmW frequency can be tuned by changing the wavelength of the local optical oscillator or the frequency of the subcarriers. When the channel spacing is 25 GHz, direct generation of ~26-~60 GHz can be realized. This technique can be applied to the channel spacing of 12.5 GHz [5], which was verified experimentally in [6].

An alternative technique for the generation of mmW in a RoF downlink is to use a remote RF carrier [7]. It is assumed that the DWDM has a channel spacing of 25 GHz. Each optical carrier carries four microwave subcarriers at 7, 10, 13 and 16 GHz and an electrical microwave tone at 23 GHz. At each BS, 50% of the optical carrier and its all subcarriers are extracted by an optical filter, which is a FBG in the experiment. After photodection, the four microwave carriers at 7, 10, 13 and 16 GHz are obtained. Then, a second optical filter with the help of an optical circulator is used to extract the remaining 50% of the optical carrier and the electrical microwave tone at 23 GHz. After photodetection, a microwave tone of 23 GHz (LO) is generated locally. With electrical mixing, the four RF carriers are up-converted to 30, 33, 36 and 39 GHz. If a further frequency up-conversion is required, an additional local electrical oscillator with the appropriate frequency difference of around several GHz should be used.

A third technique for generating mmW at BSs is based on phase modulation and direct photodetection [8-9]. Light is modulated with a phase modulator at frequency of $f_{RF}/2$. Thus two optical subcarriers at $\pm f_{RF}/2$ are created around the optical carrier. The optical carrier is removed by an optical notch filter and the two subcarriers, separated by f_{RF} are used as two optical carriers. The two optical carriers are modulated by a baseband signal. Thus, the same data signal is imposed onto the two optical carriers. By photodetection of the two optical carriers, an RF signal at f_{RF} is generated. Because of chromatic dispersion induced walk off between the optical carriers, the photodetected RF signal will suffer from power fading. In addition, the DWDM channel spacing must be greater than f_{RF}. The phase modulator can be replaced by a MZM with carrier-suppressed modulation, but it will suffer from the bias drifting problem, leading to poor system stability.

Now we consider mmW reception at BSs. For the uplink, an antenna receives an mmW signal to be sent to the CS. To directly use DWDM multiplexers and demultiplexers, one may implement photonic frequency down conversion of the mmW signal before optical demultiplexing. A simple technique was recently proposed by us [5], in which two optical sources, phase-locked and polarization-aligned, are applied to an EAM, driven by an mmW signal at f_{RF}: Since DSB SCM is used, each light has two subcarriers. Considering the fact that the mmW frequency f_{RF} is very close to the frequency separation Δf, of the two optical sources, one source is only a few GHz away from the USB subcarrier of the other. Thus, by optical filtering, frequency down conversion is obtained by changing the optical carrier. A phase-locked and polarization-aligned optical source can be obtained using supercontinuum sources [6].

A second technique for implementing mmW frequency down conversion is to use optical carrier suppression [10]. Recently, this technique was simplified by using SCM [11]. When an MZM is used for frequency shifting or frequency down conversion, SSB SCM can be generated. Similarly, an EAM can also be used to perform frequency shift, but it can only generate DSB SCM. In addition, MZMs are polarization sensitive which are not suitable for DWDM RoF systems. On the other hand, EAMs can be polarization insensitive which make them ideally suited for DWDM RoF systems.

A third technique for RoF uplink transmission is to transmit the light from a remote optical source located at the CS to a BS, which is used as the uplink optical carrier [12]. In addition, the remote light carries an mmW tone. The detected mmW tone is then mixed with the uplink RF carrier to generate a radio signal with an intermediate frequency. Thus frequency interleaving does not occur and current DWDM multiplexers and demultiplexers can be used for the uplink.

D. Nonlinear Distortion and Its Suppression

Due to the nonlinearities of the laser or the external modulator and other inline optical components, harmonic distortions (HDs) and inter-modulation distortions (IMDs) are generated in a SCM RoF system. A technique to tackle the nonlinear distortions is to use a small optical modulation index (OMI). In addition, nonlinear distortions can be suppressed by using some other techniques, such as the use of a balanced system which can suppress second-order HDs and IMDs [13]. Other techniques to reduce modulation-induced nonlinear distortions include the technique of pre-distortion [14] and the use of a linearized modulator [15].

III. OPTICAL GENERATION OF MMW SIGNALS

Extensive investigations of optical microwave generation systems have led to a variety of electrical signal generation methods. These methods employ the techniques such as automatic frequency control, optical injection locking, optical phase locking, and optical external modulation. The short- and long-term frequency stability of the optically generated mm-wave signals is critical because the resulting signal is generally used either as a microwave carrier or a local oscillator signal in system applications. Optical frequency locking, optical injection locking and optical phase lock loop (OPLL) have all been used to improve the long-term frequency stability. An OPLL, when used in combination with a narrow linewidth optical source, can generate an mmW signal with high short-term frequency stability. The short-term stability is directly related to the spectral purity of the generated signal and it can be characterized with phase noise measurement techniques. In this section, we will introduce some novel techniques to generate optically mmW signals for mmW photonics applications.

Phase Locking of Two Ultra-Quiet Lasers

The first technique is related to the use of phase locking of two ultra-quiet lasers, in which stringent requirement for a very short feedback loop is considerably relaxed [16-22]. A line-width as low as a few kHz with an exceptional long-term wavelength stability has been demonstrated for a semiconductor laser with an external FBG by the presence of a length of saturable absorber (SA) within the external cavity. Spectral line narrowing in the SA doped-fiber external-cavity laser (DFECL) has been observed experimentally.

We have generated a heterodyne beat note at a photodetector using two DFECLs by coupling their outputs using a fiber coupler. Good electrical spectrum purity is observed. The advantage of such a scheme is the considerable increase in the feed-back loop time constant and hence a reduction in the control-loop frequency, as a result of the doped-fiber lifetime. As the self-induced dynamic grating in the SA can only change slowly, the effect on the lasers linewidth and frequency stability is profound, the result being that the requirement for the PLL feedback time may be reduced to a few times the upper-state relaxation frequency.

MmW Generation Using External Modulation Techniques

We have also demonstrated two alternative approaches using an intensity modulator (IM) and a phase modulator (PM) for the generation of a tunable mmW source with high system stability and low phase noise.

In the first approach [23], continuously tunable mmW signal is generated without using a tunable optical filter. The system consists of an optical intensity modulator which is biased to suppress the odd-order optical sidebands. A narrow band FBG serving as a fixed wavelength notch filter is then used to filter out the optical carrier. A stable, low-phase noise mmW signal that has four times the frequency of the electrical drive signal is generated at the output of a photodetector. A 32 to 50 GHz mmW signal is observed on an electrical spectrum analyzer when the electrical drive signal is tuned from 8 to 12.5 GHz. The quality of the generated mmW signal is maintained after transmission over a 25-km single mode fiber.

However, to suppress the even-order optical sidebands, the

optical IM should be biased to operate in the nonlinear region, which can cause problems due to bias drift, leading to poor system robustness. A solution to this problem is to use an optical PM since no bias adjustment is required [24]. In the second approach, a tunable laser is used as the optical source, which is sent to the PM through a polarization controller. The wavelength of the optical carrier is set to match the maximum attenuation wavelength of the FBG notch filter. A microwave signal source tunable from 12.5 to 25 GHz is applied to the PM. Optical sidebands at the output of the FBG filter are amplified by an erbium-doped fiber amplifier (EDFA), and then transmitted over 60 km of single mode fiber. The beating of these optical sidebands at a photodetector generates the required mmW signals. To maintain the phase relationship of the phase modulated signal at a remote site, chromatic dispersion compensation should be employed. In the system, we use a length of dispersion compensating fiber to compensate for the chromatic dispersion.

Other Techniques for MmW Generation

We have recently demonstrated three other techniques to generate mmW signals. One technique is to use a dual wavelength single-longitudinal-mode fiber ring laser [25]. An SOA is employed as the gain medium in the ring cavity. The advantage of using an SOA instead of an EDFA is that the SOA has less homogeneous line broadening, which makes it possible to generate stable two wavelengths with small spacing at room temperature. In the proposed configuration, an ultra narrow dual-transmission-band FBG in combination with a regular FBG is used to ensure the generation of two wavelengths that are single longitudinal mode. Since the two lasing wavelengths share the same gain cavity, the relative phase fluctuations between the two wavelengths are low and can be used to generate a low phase noise microwave signal without the need of a phase lock loop and a microwave reference source. Three dual-wavelength ultranarrow transmission-band FBGs with wavelength spacings of 0.148, 0.33, and 0.053 nm are incorporated into the laser cavity. Microwave signals at 18.68, 40.95, and 6.95 GHz are respectively obtained by beating the dual wavelengths at a photodetector. The spectral width of the generated microwave signals as small as 80 kHz with frequency stability better than 1 MHz in the free-running mode at room temperature is obtained.

Another technique to generate an mmW signal is to use a rational-harmonic, actively mode-locked fiber ring laser [26]. The microwave signal is generated by beating the actively mode-locked longitudinal modes from the rational mode-locked fiber ring laser at a photodetector. The phases of the longitudinal modes are locked, which ensures a generated microwave signal with very low phase noise. In the proposed approach, the generated microwave signal has a frequency a few times higher than the microwave drive signal. Therefore, only a low-frequency reference source and a low-speed modulator are required. A rational-harmonic, actively mode-locked fiber ring laser is experimentally demonstrated. With a microwave drive signal at 5.52 GHz, a microwave signal with a frequency that is four times the frequency of the microwave drive signal at 22.08

GHz is generated. The generated microwave signal is very stable with a spectral width of less than 1 Hz.

The third technique to generate an mmW signal is based on FWM in an SOA [27]. In the proposed system, two phase-correlated optical wavelengths generated by a discriminator-added optical phase-locked loop (OPLL) [28] are used as two pumps to generate the FWM process in the SOA. Two idlers with a wavelength spacing of three times that of the two pump wavelengths are obtained. The two pump wavelengths are then removed by two cascaded FBGs serving as an optical notch filter. By beating the two idlers at a photodetector, an mmW signal with a frequency that is three times that of the OPLL reference source is generated. The key advantage of this technique is that it only uses a low frequency microwave reference.

IV. ALL-OPTICAL MMW SIGNAL PROCESSING

The key function of a RoF network is to distribute microwave and mmW signals over optical fiber, to take the advantages of the low loss, low dispersion and large bandwidth of optical fiber links. On the other hand, it is also highly desirable that the distributed signals be processed directly in the fiber link without extra optical to electrical and electrical to optical conversions. In this section, some recent advances in all-optical microwave signal processing, including all-optical microwave filtering and all-optical microwave mixing for RoF networks, will be reviewed.

A. All-Optical Microwave Filtering

All-optical microwave filters can be easily implemented with high Q, large tunability and reconfigurability. The key difficulty involving the processing of microwave signals in the optical domain is that photodetection must be incoherent in order to avoid optical interference, which leads to system instability. The use of incoherent detection limits the coefficients of the optical delay line filters to be all positive, leading to lowpass filtering only. For RoF applications, bandpass filtering is highly desirable. We have proposed three new techniques recently to implement all-optical microwave filtering with bandpass functionality.

Our first technique to implement an optical microwave bandpass filter is to use an optical phase modulator in conjunction with a length of dispersive fiber [29-30]. The phase modulation to intensity modulation (PM-IM) conversion is realized in the dispersive fiber. The PM-IM conversion has a transfer function with a notch at dc. The baseband resonance of a typical lowpass filter is eliminated with the help of the notch. Although it is a bandpass filter, in this approach it has no negative coefficients; therefore the filter has relatively high sidelobes.

In the second scheme, an optical microwave bandpass filter with negative coefficients has been demonstrated. The negative coefficients are generated through PM-IM conversion, by locating the optical carrier wavelengths at the quadrature frequencies of the positive or the negative slopes of an optical filter [31]. PM-IM conversions with microwave signals that are out of phase are generated, leading to the generation of negative coefficients. A two-tap bandpass filter with one negative tap was experimentally demonstrated. The tunability of the proposed filter was also investigated. The proposed approach has the potential to implement all-optical multi-tap microwave filters with arbitrary number of positive and negative coefficients, to ensure a bandpass filter with low sidelobes and a flat-top transmission band.

The key problem with this approach is that the system is not stable because the optical filter is sensitive to the environmental changes such as temperature and variations To solve this problem, in a third scheme, the PM-IM conversion is performed using a chirped FBG [32]. Positive and negative coefficients are obtained through conversion from phase modulation to intensity modulation by passing the phase-modulated optical carriers through chirped FBGs having group-delay responses with positive and negative slopes. A two-tap transversal microwave filter with one negative coefficient was thus experimentally demonstrated.

B. All-Optical Microwave Mixing

In addition to all-optical microwave filtering, all-optical mixing is another import topic in all-optical microwave signal processing. Recently, we have proposed, for the first time, an approach to achieve both all-optical microwave mixing and filtering in an all-optical processor [33]. The approach is based on an electro-optic phase modulator and a length of single mode fiber. The first function of the phase modulator is to perform all-optical microwave mixing. The mixed signals at the output of the phase modulator are then fed to the single mode fiber, which acts as a dispersive device for bandpass filtering, and distributes the mixed signal to a remote site. The combination of the phase modulator, a multiwavelength laser source and the single mode fiber link forms an all-optical microwave bandpass filter, which can be designed to have a passband located at the up- or down-converted microwave frequency. Frequency components other than the up-converted or down-converted frequency component will be rejected by the bandpass filter.

Recently, the electrooptic phase modulator and a length of single mode fiber were demonstrated [34]. The function of the single mode fiber is the same as that in [33], to serve as a dispersive device at the same time as a transmission medium. The input to the phase modulator is a DSB optical signal with a low subcarrier frequency. The frequency mixing is implemented at the phase modulator, to which a local oscillator frequency is applied. A subcarrier frequency conversion from 3 to 11.5 GHz performed over a 25-km radio-over-fiber link is experimentally demonstrated.

V. Low-Cost Substrate Integrated Circuits for mmW RoF Applications

One of the critical hurdles in the development of microwave and mmW photonic systems is related to the design and integration of electro-optical devices. Ideally, these should be broadband and fully compatible with other mmW front-end components including transmitting and receiving antennas.

Generally, traveling-wave devices are used to provide phase velocity matching conditions between electrical and optical signals that are able to offer such broadband characteristics. They are designed in the form of microstrip and coplanar waveguides on top of multilayered electro-optical substrates. In this case, other basic requirements for traveling-wave structures are the provision of low-loss microwave transmission and also efficient overlap of electrical and optical fields within the active region of the guided-wave regions.

We have proposed and developed the concept of substrate integrated circuits (SICs) [35]-[45], which can be effectively used for both electrical and optical waveguide structures. The basic principle of the SICs is to design or synthesize the usually conventional non-planar structures in the form of "planar circuits" so that the planar and "non-planar" structures can be made onto single substrates at low cost with the existing processing or manufacturing techniques. Of course, low-cost mmW circuits and systems can also be made in multilayer platforms in which planar and "non-planar structures" are fully integrated in a three-dimensional manner. The "non-planar" and planar integrated structures can be regarded as "volume-field circuits" and "surface-field circuits", respectively. A very large number of microwave and mmW SICs have been demonstrated with low-cost fabrication techniques, including passive and active components as well as antennas at microwave and mmW frequencies.

In our work, we have in principal examined the possibility of developing a system-on-substrate approach that allows the integration of complete mmW front-end blocks together with traveling-wave electro-optical devices on the same multilayered electro-optical substrate. In our proposed electro-optical device integration with mmW front-ends, the SIC structures are effectively combined with the traveling-wave topologies with inherent electro-optical conversion or modulation and demodulation.

VI. Conclusions

In this paper, we have reviewed a wide range of techniques for the development of innovative and maintenance-free RoF networks from device design to subsystems and system architecture. In particular, we have reviewed several new architectures for RoF systems based on DWDM technology. Techniques for the generation and processing of microwave and mmW signals in the optical domain were also reviewed. Theoretical and experimental results were obtained that have effectively validated our proposed schemes. One challenging issue is to generate a high-level RF/mmW power at BSs through optical means, which is critical for practical implementation of maintenance-free RoF systems. Additionally, the potential of using SICs to implement system-on-substrate or system-on-chip for future mmW RoF applications were also discussed.

References

[1] S. Blais and J. P. Yao, "Optical single sideband modulation using an ultra-narrow dual-transmission-band fiber Bragg grating," *IEEE Photon. Technol. Lett.*, to be published.

[2] G. Smith, D. Novak, and Z. Ahmed, "Overcoming chromatic dispersion effects in fiber-wireless systems incorporating external modulators," *IEEE Trans. Microwave Theory Tech.*, vol. 45, pp.1410-1415, 1997.

[3] A. Narasimha, X. Meng, M. Wu, and E. Yablonovitch, "Tandem single-side band modulation scheme for doubling spectral efficiency of analogue fiber links," *Electron. Lett.*, no. 13, vol. 36, pp.1135-1136, 2000.

[4] H. Toda, T. Yamashita, T. Kuri, and K. Kitayama, "Demultiplexing using an arrayed waveguide grating for frequency interleaved DWDM millimeter-wave radio on fiber systems," *J. Lightwave Technol.*, vol. 21, no. 8, pp. 1735-1741, 2003.

[5] X. Zhang, B. Liu, J. P. Yao, K. Wu, and R. Kashyap, "A novel millimeter-wave band radio over fiber system with dense wavelength division multiplexing bus architecture," *IEEE Trans. Microwave Theory Tech.*, vol. 54, no. 2, pp. 929-937, 2006.

[6] T. Nakasyotani, H. Toda, T. Kuri, and K. Kitayama, "Wavelength-division multiplexed millimeter-wave band radio on fiber system using a supercontinuum light source," *J. Lightwave Technol.*, vol. 24, no. 1, pp. 404-410, 2006.

[7] C. Lim, A. Nirmalathas, M. Attyalle, D. Novak, and R. Waterhouse, "On the merging of millimeter-wave fiber radio backbone with 25-GHz WDM ring networks," *J. Lightwave Technology*, vol. 21, no. 10, pp. 2203-2210, 2003.

[8] J. Yu, G. Chang, Z. Jia, L. Yi, Y. Su, and T. Wang, "A RoF downstream link with optical mm-wave generation using optical phase modulator for providing broadband optical wireless access service," *OFC 2006, Paper OFM3, CA.*

[9] G. Chang, J. Yu, Z. Jia, and J. Yu, "Novel optical wireless access network architecture for simultaneously providing broadband wireless and wired services," *OFC 2006, Paper OFM1, CA.*

[10] T. Kuri, H. Toda, and K. Kitayama, "Dense wavelength division multiplexing millimeter wave band radio on fiber signal transmission with photonic downconversion," *J. Lightwave Technol.*, vol. 21, no. 6, pp. 1510-1517, 2003.

[11] G. Zhou, X. Zhang, J. Yao, K. Wu, and R. Kashyap, "A novel photonic frequency down-shifting technique for millimeter-wave band radio over fiber systems," *IEEE Photonics Technol. Lett.*, vol. 17, pp. 1728-1730, 2005.

[12] A. Kaszubowska, L. Hu, and L. Barry, "Remote downconversion with wavelength reuse for the radio/fiber uplink connection," *IEEE Photon. Technol. Lett.*, vol. 18, no. 4, pp. 562-564, 2006.

[13] B. Masella and X. Zhang, "A novel single wavelength balanced system for radio over fiber links," *IEEE Photon. Technol. Lett.*, vol. 18, no. 8, pp. 301-303, 2006.

[14] S. Tanaka, N. Taguchi, T. Tsuneto, and Y. Atsumi, "A predistortion-type equi-path linearizer design for radio over fiber system," *IEEE Trans. Microwave Theory and Techniques*, vol. 54, pp. 938-944, 2006.

[15] H. Tazawa and W. Steier, "Linearity and ultra-linearization of ring resonator based modulators for sub-octave bandpass analog optical links," *OFC 2006, Paper JThB23, CA.*

[16] I. A. Kostko and R. Kashyap, "Dynamics of ultimate spectral narrowing in a semiconductor fiber-grating laser with an intra-cavity saturable absorber," *Opt. Express*, vol. 14, pp. 2706-2714, Apr. 2006.

[17] F. N. Timofeev and R. Kashyap, "High-power, ultra-stable, single-frequency operation of a long, doped-fiber external-cavity, grating-semiconductor laser," *Opt. Express*, vol. 11, pp. 515-520, Mar. 2003.

[18] R. Liu, I. Kostko, K. Wu, and R. Kashyap, "Optical generation of microwave signal by doped fiber external cavity semiconductor laser for radio-over-fiber transmission," *Proc. SPIE, Photonic Applications in Nonlinear Optics, Nanophotonics, and Microwave Photonics*, R. A. Morandotti, H. E. Ruda, J. P. Yao, Eds., vol. 5971, pp. 455-462, Sep. 2005.

[19] R. N. Liu, I. A. Kostko, R. Kashyap, K. Wu, and P. Kiiveri, "Inband-pumped, broadband bleaching of absorption and refractive index changes in erbium doped fiber," *Opt. Commun.* vol. 255, pp. 65-71, Nov. 2005.

[20] I. A. Kostko and R. Kashyap, "Dynamics of ultimate spectral narrowing in a semiconductor fiber-grating laser with an intra-cavity saturable absorber", Opt. Express, vol. 14, no. 7, pp. 2706-2714. Apr. 2006.

[21] R. Liu, I. A. Kostko, K. Wu, and R. Kashyap, "Side mode suppression using a doped fiber in a long external- cavity semiconductor laser operating at 1490 nm," Opt. Express, vol. 14, no. 20, pp. 9042-9050, Oct. 2006.

[22] R. Liu, I. Kostko, K. Wu, and R. Kashyap, "Tuning characteristics of erbium doped fiber long external cavity semiconductor laser for radio-over-fibre applications," *Proc. SPIE, Photonics North*, Quebec, Canada, June 2006.

[23] G. Qi, J. P. Yao, J. Seregelyi, C. Bélisle, and S. Paquet, "Generation and distribution of a wideband continuously tunable mm-wave signal with an optical external modulation technique," *IEEE Trans. Microwave Theory Tech.*, vol. 53, no.10, pp. 3090- 3097, Oct. 2005.

[24] G. Qi, J. P. Yao, J. Seregelyi, C. Bélisle, and S. Paquet, "Optical generation and distribution of continuously tunable millimeter-wave signals using an optical phase modulator," *J. Lightwave Tech.*, vol. 23, no.9, pp. 2687-2695, Sep. 2005.

[25] X. Chen, Z. Deng, and J. P. Yao, "Photonic generation of microwave signal using a dual-wavelength single-longitudinal-mode fiber ring laser," *IEEE Trans. Microwave Theory Tech.*, vol. 54, no.2, pp. 804-809, Feb. 2006.

[26] Z. Deng and J. P. Yao, "Photonic generation of microwave signal using a rational harmonic mode locked fiber ring laser," *IEEE Trans. Microwave Theory Tech.*, vol. 54, no.2, pp. 763-767, Feb. 2006.

[27] Q. Wang, H. Rideout, F. Zeng, and J. P. Yao, "Millimeter-wave frequency tripling based on four-wave mixing in a semiconductor optical amplifier," *IEEE Photon. Technol. Lett.*, to be published.

[28] H. Rideout, J. Seregelyi, S. Paquet, and J. P. Yao, "Discriminator-aided optical phase-lock loop incorporating a frequency down-conversion module," *IEEE Photon. Technol. Lett.*, to be published.

[29] F. Zeng and J. P. Yao, "All-optical bandpass microwave filter based on an electro-optic phase modulator," *Opt. Express*, vol. 12, no. 16, pp. 3814-3819, Aug. 2004.

[30] F. Zeng and J. P. Yao, "Investigation of phase modulator based all-optical bandpass filter," *IEEE J. Lightwave Technol.*, vol. 23, no. 4, pp. 1721-1728, Apr. 2005.

[31] J. Wang, F. Zeng, and J. P. Yao, "All-optical microwave bandpass filters with negative coefficients based on PM-IM conversion," *IEEE Photon. Technol. Lett.*, vol. 17, no.10, pp. 2176- 21780, Oct. 2005.

[32] F. Zeng, J. Wang, and J. P. Yao, "All-optical microwave bandpass filter with negative coefficients based on an electro-optic phase modulator and linearly chirped fiber Bragg gratings," *Opt. Lett.*, vol. 30, no. 17, pp. 2203-2205, Sept. 2005.

[33] F. Zeng and J. P. Yao, "All-optical microwave mixing and bandpass filtering in a radio-over-fiber link," *IEEE Photon. Technol. Lett.*, vol. 17, no. 4, pp. 899-901, Apr. 2005.

[34] J. P. Yao, G. Maury, Y. L. Guennec, and B. Cabon, "All-optical subcarrier frequency conversion using an electrooptic phase modulator," *IEEE Photon. Technol. Lett.*, vol. 17, no.11, pp. 2427-2429, Nov. 2005.

[35] K. Wu, "Integration and interconnect techniques of planar and non-planar structures for microwave and millimeter-wave circuits – current status and future trend", *Proceeding of Asia-Pacific Microwave Conference*, Taipei 2001, pp. 411-416.

[36] K. Wu and F. Boone, "Guided-wave properties of synthesized non-radiating dielectric waveguide for substrate integrated circuits (SICs)", *2001 IEEE MTT-S Inter. Microwave Symp.*, pp. 723-726, Phoenix, USA.

[37] K. Wu, D. Deslandes, and Y. Cassivi, "The substrate integrated circuits – a new concept for high-frequency electronics and optoelectronics," *TELSIKS'03*, Nis, Serbia and Montenegro, pp. P-III to P-X.

[38] K. Wu, "Towards system-on-substrate approach for future millimeter-wave and photonic wireless applications," *Inter. Joint Conf. Proc. of MINT-MIS/TSMMW*, pp. 229-232, Feb. 2005, Seoul.

[39] D. Deslandes and K. Wu, "Single-substrate integration techniques for planar circuits and waveguide filters," *IEEE Trans. Microwave Theory Tech.*, vol. 51, pp. 593-596, Feb. 2003.

[40] F. Xu and K. Wu, "Guided-wave and leakage characteristics of substrate integrated waveguides," *IEEE Trans. Microwave Theory Tech.*, vol. 53, pp. 66-73, Jan. 2005.

[41] W. D'Orazio, K. Wu and J. Helszajn, "A substrate integrated waveguide degree-2 circulator," *IEEE Microw. Wirel. Compon. Lett.*, vol. 14, no. 5, pp. 207-209, May 2004.

[42] Y. Cassivi and K. Wu, "Substrate integrated non-radiative dielectric (SINRD) waveguide", *IEEE Microw. Wirel. Compon. Lett.*, vol. 14, no. 3, pp. 89-91, Mar. 2004.

[43] J. Dallaire, K. Wu, "Complete characterization of transmission losses in generalized non-radiative dielectric (NRD) waveguide," *IEEE Trans. Microwave Theory Tech.*, vol. 48, pp. 121-125, Mar. 2000.

[44] D. Deslandes and K. Wu, "Accurate modeling, wave mechanisms, and design considerations of substrate integrated waveguide," *IEEE Trans. Microwave Theory Tech.*, vol. 54, no. 6, pp. 2516-2526, June 2006.

[45] A. Patrovsky and K. Wu, "Substrate integrated image guide (SIIG)—a novel planar dielectric waveguide technology for millimeter-wave applications," *IEEE Trans. Microwave Theory Tech.*, vol. 54, no. 6, pp. 2872-2879, June 2006.

Theoretical and Experimental Investigations of Lightcraft/ Impulsar Related Problems

V.V. Apollonov

A. M. Prokhorov General Physics Institute, Vavilov str.38, 119991, Moscow, Russia

I. INTRODUCTION

The objective of the project is to accomplish a circle of experimental, engineering and technological works on creation of a high efficiency laser rocket engine. According to estimations done by specialists, the capabilities of the commercial launch market in 2007 will rise 40% compared to 2004. Taking this into account, developed countries are conducting investigations on the feasibility of rocket engine alternatives to the existing ones, which operate on chemical fuel. Prospective rocket systems of this new concept should be classified as laser rocket engines. It is implied that the creation of a spaceship where the initial part (acceleration) of its trajectory will obtain thrust by means of a continuous sequence of laser pulses directed to it from the Earth's surface.

The importance of this problem lies in the fact that laser based systems are essentially more economic than conventional jet engines that work with chemical fuel. At the initial stage of the flight, these will carry a small amount of easily sublimated matter stored aboard. In this case, the specific cost of load transportation to space may drop by $200-$500 USD/kg i.e. approximately; one hundred times less than contemporary costs. Also, the possibility of orbitparameter correction by means of the same laser system intended for launch is particularly attractive.

At present, works on investigating the feasibility to create a laser propulsion engine is in the making. In the USA, within the frame of the "Lightcraft" project, a laser engine has been developed. Thus, in November of 2000 the company "Lightcraft Technologies" successfully conducted tests of a model rocket, which reached a height of 70 m at 12.7 sec under the action of a jet initiated by a high power laser pulse. This experiment exploited the use of an old fashion pulsed 10 kW CO_2 laser, with a long pulse duration and low repetition rate. Propulsion resulted from mass blow-away of a special polymer matter from the concave surface in the bottom of the rocket where the laser beam was directed.

As early as 1973 at the P. N. Lebedev Physical Institute (USSR Academy of Sciences) under leadership of acad. A. M. Prokhorov, works on the investigation of laser propulsion engine feasibility were investigated [4]. The engine setup operated with a laser beam directed onto a reflector at its rear. The reflector concentrated the radiation in the air thus providing optical breakdown (microexplosion),which in turn created thrust. Successful results were obtained by testing various reflector designs, which simultaneously served as receivers of the compression wave to provide thrust. It should be noted that all the experiments described above were carried out with electric discharge CO_2 lasers of low power (below 10 kW). However, to launch "hi-tech" equipment into orbit (communications, Internet, photomonitoring) requires very high radiation power. E.g., to launch a 300 kg satellite (more than 85% of commercial launches), a laser of at least 10 MW is required. At present, such a laser may only be of the gasdynamic type, because only for this type of laser has the technology developed enough; and has enabled to formulate such a problem. Besides, the laser must operate in pulse-periodic mode with a high repetition rate of short pulses to rule out screening of the radiation by plasma formation at the engine operation, and also for boosting its operation.

In the opinion of specialists (classic rocket scientists), such laser-based systems may find application in cheap single-stage means of launching satellites (nano-,micro- and mini-) with total weight in the gap of 5 - 300 kg. At the first stage of flight of such an apparatus, up to a height of 30-50 km, it is supposed to utilize atmospheric air as the working body. Beyond that height, until orbit is reached, an on board supply of fuel may be used, whose mass should not exceed 10 - 15% of the apparatus's weight.

Experience on creation of high-power gas-dynamic lasers (GDL) has accumulated at A.M. Prokhorov General Physics Institute (RAS), NPO "Energomash", "KBKhA" and "M. V. Keldysh Center". In the last years, JSC "Energomashtekhnika" has successfully conducted experimental investigations on the development of pulse-periodical mode [3], high-power and well-refined CW laser systems. On the other hand, NPO "Energomash", NPO "Molniya" and "KBKhA" for many years have also developed liquid fuel rocket jets. Their highefficiency designs are applied in most Russian and former USSR cosmic rockets.

The Institute of Laser Physics and the Institute for Theoretical and Applied Mechanics (both from RAS, Siberian branch) have also made valuable contributions to this field. This encourages the cooperation above to initiate the experimental realization of the laser rocket engine and the creation of a system for the launch of mini-satellites, for the purpose of constructing a global system of state-of-the-art wide band communications with laser beams and super-high speed channels for the Internet. This cumulative work project is expected to be a prominent step to future launches of super-light satellites at low orbits and even transportation into space piloted spacecrafts. Realization of the project will give capabilities to create highly economic laser jets. Due to the possibility of their multiple uses and dropping the specific costs of payload transportation into space, an opportunity arises to decrease the entire cost of rockets and to transport general-purpose payloads into space.

The main advantage of the new approach is associated with the fact that the source of energy of propulsion and the load are spatially disconnected and the start weight of the rocket may be dropped from 705 tons ("Proton") to a few tons of payload only. Earlier the great K. Tsiolkovski prophesied that all launches of future rockets would be accomplished by electromagnetic waves directed from a distant power source - The laser had not invented at that time.

The highest interest of scientists in our days is related to the successful solution of the creation of high-power laser sources with high repetition frequency (50-100 kHz) of pulses and small duration (100-200 ns). Such a mode is developed and realized on a basis of high-power CO_2 based GDL [3,6] and may be transferred to other types of high-power lasers like HF/DF, Nd: YAG with solid-state pumping and COIL. At present, a 10-20 MW laser project with controlled temporal structure of radiation is under stage of active preparations. Also it should be noted that works in this area, considering great perspectives of various applications, are in progress in USA, Germany, Japan, Great Britain, France, China and some other countries.

Meanwhile, the participants of these works emphasize pronounced status of GDL as of the most prospective system from the point of view of scalability to a few tens of MW level and other parameters important for such an applications. Works on refinement of scientific, engineering and technological solutions with respect to GDL were not discontinued till today. The transformation of pulsed periodic laser radiation into a quasistationary wave (QSW) under the conditions of an optical pulsating discharge (OPD) for the case of lightcraft engine was studied for the first time in our publications [1,2].

Unlike a CW optical discharge, no physical constraints are imposed on the velocity of OPD propagation in a gas. In contrast to a single laser spark, the shock waves (SW) generated by an OPD merge to form a QSW propagating in a preferred direction in the surrounding gas. About 35% of the laser power is transformed into SW by an OPD, which may exert a considerable influence on the surrounding medium.

II. MECHANISM OF MERGING SHOCK WAVES

The mechanism of SW merging can be described as follows. Periodic perturbations with an initial velocity exceeding the velocity of sound are successively generated in a continuous medium. The velocity V of propagation of the pulse region is lower than velocity of sound. The SW merge to form a QSW if the parameters of the pulses and the medium satisfy the criteria formulated in the study [2]. Depending on the space-and-time structure of the pulses, the mechanism manifests itself in the form of effects characterized by a long, elevated pressure region.

This mechanism does not impose any constraints on the type of the medium and the source of pulses, nor on its energy (which

is important for producing long waves). The QSW can move from the source, which has a point size in a preferred direction. The SW can also merge during supersonic propagation of an OPD, but the length of the combined SW in a direction perpendicular to its front is small. The present study is aimed at analyzing the interaction between an OPD and a gas medium under the conditions when the wave combination mechanism is manifested. Earlier, we had considered the situations when the OPD is stationary or moves along the laser beam axis at a constant relative velocity. The criteria for combining SW into QSW in air were determined in [2].

Let us briefly dwell on the importance of the problems being studied here. It is found experimentally that an OPD operates at any gas velocity $V = 0$ - 400 m/s if the radiation power exceeds a certain threshold value. However, the transformation of laser radiation into intense interacting SW is possible only for a certain ratio of the parameters of the OPD and the medium. Hence it is important to set the conditions for the formation of QSW in various gases and to determine whether any constraints are imposed on the spark energy. Apart from the acoustics, OPD and QSW are also of interest for solving aerospace problems. For example, it was found experimentally that aerodynamic drag decreases approximately to half its value if an OPD is initiated in front of a body in an incoming supersonic flow. Increasing the coupling factor J characterising the efficiency of laser radiation application for speeding up an aircraft is one of the main problems in the development of a laser engine. We believe that the value of J can be increased substantially by using QSW and a powerful pulse-periodic radiation of small duration (~ 0.1—1 mks) and a high pulse repetition rate (tens of kHz). Considerable advances towards the development of such lasers were made in [3] where an average power ~ 10 kW of pulse-periodic radiation was attained and the possibility of its further increase was demonstrated.

Merging of SW generated by OPD -How it can be done? In keeping with the above hypothesis, the wave combination mechanism operates in various media for all energies and all types of pulse sources. The following questions are considered in the present study: can we introduce a universal criterion of wave combination for different gases? Does the mechanism hold for arbitrary pulsation energy? Which constraints on the parameters of the pulse source (OPD) are imposed due to the requirement of nonlinearity of interaction of the source with the medium? Certain constraints were employed in the course of our investigations. We considered a stationary OPD or one propagating in a gas at a constant velocity lower than the velocity of sound. Equations of gas dynamics in the twodimensional axisymmetric approximation were solved in our publication [2]. Let us consider the QSW mode in a lightcraft engine. One of the methods for producing thrust in a laser engine can be described as follows. Pulse-periodic radiation is incident on a focusing reflector producing periodically repeated laser sparks at the focus. The sparks generate SW producing an alternating force at the reflector (compression and low-pressure phases in SW). Typical value of the coupling factor J, characterizing the efficiency of employment of laser radiation since 1973 is equal to ~ 100—500 N/MW [4,5]. Here, we suggest that a plane QSW be used for a substantial increase in the value of J. The QSW produces a high constant pressure on a large area of the reflector. A simplified scheme permitting calculations in the 2D axisymmetric approximation can be described as follows. The OPD has the shape of a disc whose plane is perpendicular to the reflector axis and whose radius is much larger than its length and smaller than the distance (~ 20—50 cm) between the OPD and the reflector. The SW's generated in the direction of the reflector merge to form a QSW in the region between the OPD and the reflector. Such an idealization might correspond to a 2D OPD matrix produced synchronously by many beams. We carried out computer simulations to estimate the coupling factor J in the case when a plane QSW is used in a lightcraft engine. The repetition rate was chosen from the conditions of formation of a plane QSW.

The simulation technique consisted in the following ideal picture. At the initial stage, the OPD burns in a free gaseous space. After the passage of several hundred microseconds, a QSW is formed in front of the OPD, in which the gas flows at a velocity ~ 300 m/s in the same direction as the OPD. After this, a wall with which the wave interacts appears on the path of the QSW. The range of the high-pressure region increases with time due to the curvature of the leading front of the QSW. It follows from our computations that $J = 2000$ N/MW for $p0 = 1$ atm can be very real even for non-optimal conditions. Thus, an OPD can be stationary or move at a high velocity in a gaseous medium. However, stable SW generation occurs only for a certain relation between the radiation intensity, laser pulse repetition rate, their filling factor, and the OPD velocity-wide spectrum of adjustable parameters. The OPD generates a QSW in the surrounding space if it is stationary or moves at a subsonic velocity and its parameters satisfy the above conditions. The mechanism of SW combination operates in various media in a wide range of pulse energies. The results of preliminary investigations show that the efficiency of the laser radiation can be increased substantially when a QSW is applied for producing thrust in a lightcraft engine.

III. The Regime of Light Supported Detonation Waves Merging

Typical velocities of LSDW reasonable for our applications: V ~ 5-15 km/s. Threshold values for free space and for near surface conditions: 10 – 100 MW/cm2 for λ=10.6mkm. Expression for velocity [2]:

$$V = \left[2 \cdot \left(\gamma^2 - 1\right)\right]^{\frac{1}{3}} \cdot \left(\frac{I}{\rho_0}\right)^{\frac{1}{3}} \approx 1830 \left(\frac{I'}{P_0}\right)^{\frac{1}{3}} \quad [\text{m/s}]$$

Where I –laser radiation intensity, $r0$- gas density, I [MW/cm2], $P0$ [atm]. The length of laser spark for LSDW for unlimited duration and for constant intensity- Z_P. Expression for Z_P:

$$Z_p \frac{d_b}{F_f} = 0.013 \left(\frac{W}{P_0}\right)^{1/2} \left\{1.93 + \ln P_0 + \ln\left[\left(\frac{W}{P_0}\right)^{1/2} \left(6.3 + \ln P_0\right)^{3/4}\right]\right\}^{3/4}$$

Where db [cm] - diameter of the beam, Ff – focal length, W [MW] –laser pulse power, P_0 [atm.] air pressure. For $P_0 = 1$ & 0.1 atm. From (1) we can get:

$$Z_p \frac{d_b}{F_f} = 0.013 \left(\frac{W}{P_0}\right)^{\frac{1}{2}} \left\{3.32 + \frac{1}{2} \ln\left(\frac{W}{P_0}\right)\right\}^{3/4} \quad (1.a)$$

$$Z_p \frac{d_b}{F_f} = 0.013 \left(\frac{W}{P_0}\right)^{\frac{1}{2}} \left\{0.67 + \frac{1}{2} \ln\left(\frac{W}{P_0}\right)\right\}^{3/4} \quad (1.b)$$

The time of laser radiation is on- t = tr and up to the moment when the regime of LSDW is over, for this case length of laser spark:

$$Z_P = 1.32 \left(\frac{F_f}{d_b}\right)^{\frac{2}{5}} W^{\frac{1}{5}} \cdot t_r^{\frac{3}{5}} \frac{1}{\left(AP_0\right)^{\frac{1}{5}}} = 0.67 \left(\frac{F_f}{d_b}\right)^{\frac{2}{5}} \left(\frac{W}{P_0}\right)^{\frac{1}{5}} \cdot t_r^{\frac{3}{5}} \quad (2)$$

Here tr [mks].

From (1) and (2) one can get the time of LSDW decay due to the laser beam expansion:

$$t_r \frac{d_b}{F_f} = 0.0014 \left(\frac{W}{P_0}\right)^{1/2} \left\{1.93 + \ln P_0 + \ln\left[\left(\frac{W}{P_0}\right)^{1/2} \left(6.3 + \ln P_0\right)^{3/4}\right]\right\}^{5/4} \quad (3)$$

For $P_0 = 1$ atm. and 0.1 atm expression (3) can be written as:

$$t_r \frac{d_b}{F_f} = 0.0014 \left(\frac{W}{P_0}\right)^{\frac{1}{2}} \left\{3.32 + \frac{1}{2} \ln\left(\frac{W}{P_0}\right)\right\}^{5/4} \quad (3a)$$

$$t_r \frac{d_b}{F_f} = 0.0014 \left(\frac{W}{P_0}\right)^{\frac{1}{2}} \left\{0.67 + \frac{1}{2} \ln\left(\frac{W}{P_0}\right)\right\}^{5/4} \quad (3b)$$

Laser energy absorbed during tr is:

$$\frac{q}{P_0} \cdot \frac{d_b}{F_f} = 0.0014 \left(\frac{W}{P_0}\right)^{3/2} \left\{1.93 + \ln P_0 + \ln\left[\left(\frac{W}{P_0}\right)^{1/2} \left(6.3 + \ln P_0\right)^{3/4}\right]\right\}^{5/4} \quad (4)$$

For Po = 0.1 и 1 atm. expression (6) can be written as :

$$q \frac{d_b}{F_f} = 0.0014 \left(\frac{W}{P_0}\right)^{\frac{3}{2}} \left\{3.32 + \frac{1}{2} \ln\left(\frac{W}{P_0}\right)\right\}^{5/4} \quad (4a)$$

$$q \frac{d_b}{F_f} = 0.0014 \left(\frac{W}{P_0}\right)^{\frac{3}{2}} \left\{0.67 + \frac{1}{2} \ln\left(\frac{W}{P_0}\right)\right\}^{5/4} \quad (4b)$$

Length of the spark- $Z_P(d_b/F_f)$, time duration and energy of pulse $t_P(d_b/F_f)$, $(q/P_0)(d_b/F_f)$ for gas pressure Po = 1, 0.1 atm can be presented by:

For Po = 1 atm	Po = 0.1 atm.	
Zp = 0.0336W0.57 Ff/df ,	Zp = 0.0667 W0.62 Ff/df	(5)
tp = 0.0068 W0.618 Ff/df,	tp = 0.01 W0.7 Ff/df	(6)

$q = 0.0068\ W1.62\ Ff/df,$ $q = 0.01\ W1.7\ Ff/df.$ (7)

Estimations for the case of well-known experiments [5] with the following parameters:
q = 280 J, tr ~ 30 mks, W(peak)~10 MW.
Real values for this boundary condition are:
1. For Po = 0.1 atm. - Zp = 0.28 cm, tr = 0.05 mks, q = 0.5 J;
2. For Po =1 atm. - Zp = 0.12 cm, tr = 0.028 mks, q = 0.283 J.
   (8)

One can see from here that the W~10 MW regime of LSDW was not effective. In that case the absorption process had the next stages: optical breakdown – LSDW- SbRW- bremsstrahlung absorption in expanding plasma.

In our case we need to find the conditions: W и tr, optimal for the energy absorption. From (5) – (7) one can get:

For Po= 1 atm.　　　For Po = 0.1 atm.
W = 21.7 (q df/Ff)0.617　　W = 15 (q df/Ff)0.588
tr = 0.0455 (q Ff/df)0.381　tr=0,0665 q0.411 (Ff/df)0.588   (9)
Zp = 0.194q0.352 (Ff/df)0.648　Zp = 0,357 q0.365 (Ff/df)0.635

All that is valid ,if OPD conditions are corresponding to:

$2{,}5 \cdot M0 > f' > 5{,}88 \cdot (1 - M0)^{1.5}$
Here: Mo = Vo/Co– gas velocity in comparison with regime of OPD/5/ (Co ≈ 340 m/s – velocity of sound in air) and pulse frequency f' = f·td , td = Rd /Co – dynamic time , Rd = (Qr / Po)1/3 [m] – dynamic radius, Qr [J] – energy of spark, Po [pa] – gas pressure. Reasonable values of Mo ~ 0.65 – 0.95.

q, J		10	100	1000	280 experiments[5]
Po = 0.1 atm.	tr, mks	0,17	0,439	1,13	0,665
	W, MW	58,1	225	873	411
	Zp, cm	0,83	1,97	4,52	2,86
Po = 1 atm.	tr, mks	0,11	0,263	0,63	0,39
	W, MW	90	372	1540	702
	Zp, cm	0,44	0,98	2,21	1,41

IV. OPTIMAL CONDITIONS FOR QCW MODE

For W $_c$ –fixed one can find the area of {Qr, f}, where the optimal conditions for parameters [5] can be satisfied:

f1 [kHz] < 40·Mo /(Qr/P0)1/3, Po [atm]   (10)
f2 [kHz] > 93·(1 - Mo)3/2 /(Qr/Po)1/3,   (11)
   Qr > Qr min,   (12)
   f3 [kHz] = WC/Qr.   (13)

For Mo = 0.8 and P0 = 1 атм. (2) – (3) can be presented by:
f1 [kHz] < 32 /(Qr)1/3,   (14)
f2 [kHz] > 8.3/(Qr)1/3,   (15)

Laser radiation parameters for launching: lightcraft mass-Ma ; launching is along of straight line St ~ 200 km ; height h ~ 50 km; final velocity- Va= 5 km/s.

The conservation law for our case:
$$0.5 M_a V_a^2 + M_a Gh + F_r S_t = F_a S_t \qquad (16)$$
$$0.5 M_a V_a^2 / M_a Gh \approx 25$$
$$F_t(t) = \int_0^{R_s} \left(\frac{P_r(t,r)}{P_0} - 1 \right) P_0 r dr$$

During the time Δt for lightcraft we have:

$$\Delta P_a = M_a \Delta V_a = \int_0^{\Delta t} F_t dt = tf \int_{t_i}^{t_{i+1}} F_t dt = tf \delta P_a = t F_a \qquad (17)$$

ΔPa = δPa f t = Fat = Jк·q·f·t = Jк·Wc·t. Here Wc – average power of pulse-periodic laser,Fa = Jr·Wc

By using Fa = Jr·Wc from (1) one can get main parameters of the task until the lightcraft approaches a velocity-Va.

$$\frac{M_a}{W_c} = \frac{J_{rc} S_t}{0.5 V_a^2 + Gh + G_r S_t} \approx \frac{2 J_{rc} S_t}{V_a^2} \qquad (18)$$

$$\frac{M_a}{W_c} \approx \frac{J_{rc} \cdot t_a}{V_a} \quad [kg/MW] \qquad (19)$$

$$t_a = \left(\frac{2 M_a S_t}{J_{cr} W_c} \right)^{\frac{1}{2}} \qquad (20)$$

Here: ta – time of acceleration, St –distance of acceleration Va –final velocity, Jrc ≈ (Jr0 + Jr1)/2 (Jr0 – for the begining and Jr1 for the height-h).Average value: Jrc ≈ 0.75 Jr0 (Jr0 = 2000 H/MW), in case of acceleration of lightcraft by QSW.

Let us make some estimations for the following parameters:
St ~ 200 km, Va = 5 km/s, h = 50 km from (11) and (13) one can get:

In the case of QSW mechanism: Ma/Wc = 25 kg/MW, ta = 80s

In the case of experimental conditions [5]: Ma/Wc = 5 kg/MW, ta = 180 s

The comparison shows that the improvement factor of the launching conditions in our case goes to the value –5, which is a pessimistic estimation only; not taking further optimization into account for future launching conditions. Estimations of thrust for the case of QSW and a 10MW pulse-periodic laser system based on GDL [3,6,7] show the ability of launching earth-orbit payloads with a total weight of more than 300kg.

V. CONCLUSION

The regime of light supported detonation wave for theoretically investigated high repetition rate temporal structure of laser radiation with much higher output power has demonstrated a very high efficiency for lightcraft applications. Quasistationary wave mode based on effect of shock waves merging has been suggested and analyzed. And some estimations of applicability of that regime to the lightcraft project have been provided.

REFERENCES

[1] ApollonovV.V.,TishchenkoV.N."Kvantovaya electronika"34,12,(2004), SPIE Proceedings of GCL/HPL,vol.5777,Prague,2004.

[2] Tishchenko V.N., Apollonov V.V., Grachev G.N., Zapryagaev V.I., Gulidov A. I. ,Menshikov Ya. I. ,Smirnov A. L. , Sobolev A. V. "Kvantovaya electronika" 34,10,(2004).

[3] Apollonov V.V., Kiiko V.V., Kislov V.I., Suzdal'tsev A.G., Egorov A.B.,Kvantovaya electron.33,753 (2003) [Quantum Electron. 33, 753 (2003)].

[4] Ageev V.P., Barchukov A.I., Bunkin F.V., Kononov V.I., Prokhorov A.M., Silenok A.S.,Chapliev N.I., Kvantovaya electron. 4(12), 2501 ,(1977)

[5] Schall W.O., Bohn W.L., Eckel H.-A., Mayerhofer W., Riede W., Zeyfang E. "Lightcraft experiments in Germany",Proceeding of SPIE V.4065,p.472,April 2000.

[6] Wallace J. "Gas-dynamic laser enters pulse-periodic mode "LFW, p.17,August 2004.

[7] Apollonov V. V. "20MW GDL for Lightcraft applications"ISBEP-3, Troy NY, 2004

[8] Apollonov V.V.,Tishchenko V.N."Kvantovaya electronika"36,7,(2006)

Optical Buffering and Time Slot Interchanging Based on an Optical Crosspoint Switch Matrix

Nan Chi(1), Dexiu Huang(1), Zhuoran Wang(2), Siyuan Yu(2)

(1) Wuhan National Laboratory for Optoelectronics, Huazhong University of Science and Technology,
Wuhan 430074, China
(2) CCR Center, University of Bristol, BS8 1TR, United Kingdom.

Abstract—We experimentally demonstrate time-slot interchanging for an optical packet consisting of a 10 Gb/s payload and a 155 Mb/s label based on active vertical coupler (AVC) crosspoint switch.

I. INTRODUCTION

Optical packet switching, in which packets are switched and buffered optically, have been proposed worldwide as a way of overcoming the envisaged future limitations of electronic packet switching technology. The purpose of the optical buffer is to store and then release the data in optical format without conversion into the electrical format [1]. One attractive function of optical buffering is to realize time-slot interchanging(TSI). As an important method of switching in the time domain, TSI can rearrange the order of the time slots in a traffic stream [2]. Furthermore, TSI can reduce the blocking probability of TDM switches [3] and therefore a contention resolution scheme in optical packet switching nodes.

In this paper, we report a flexible optical packet buffering and time slot interchanging scheme by using an active vertical coupler (AVC) based optical crosspoint switch matrix (OXS). We experimentally demonstrate opto-electrical controlled packet routing, automatic power balancing and packet buffering and TSI for a 10 Gb/s payload and a label at 155 Mb/s with only one individual switch element. We demonstrated all-optical buffering up to 9 μs and TSI for 4 serial optical packets.

II. PRINCIPLE

The OXS matrix is fabricated in quaternary semiconductor multi-layers deposited on an InP substrate. The fabric concatenates 16 of switch cells in a mesh-structure (see Fig.1). Each switch cell consists of two AVCs formed on top of the two perpendicular passive waveguides. A total internal reflection mirror vertically penetrates the active waveguide layer with an angle of 45° with respect to the two couplers directions. When the AVCs are driven by carriers, meaning that the single switch cell is in ON-state, the light injected to the passive waveguide will be coupled to the active layer owing to the optical gain and the refractive index matching. This light in active layer is then reflected by the mirror to another AVC and coupled down to the passive waveguide as the output (see Fig.1(a)). The absence of

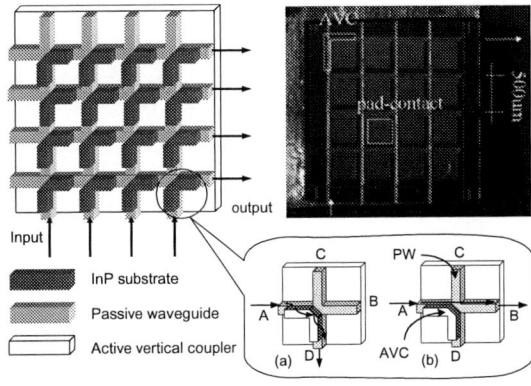

Fig.1. Schematics and the photograph of the 4x4 crosspoint switch matrix. (a) single switch cell in the ON-state, (b) single switch cell in the OFF-state

carrier injection in the active layer will lead the switch cell to the OFF-state, as illustrated in Fig.1(b). The input light will simply pass through the passive waveguide. In this way, optical signals coming from the 4 input ports will be switched to the 4 output ports by triggering ON correctly selected switch cells.

III. EXPERIMENTAL SETUP AND RESULTS

The experimental setup is shown in Fig. 2. The signal source is a wavelength tunable external cavity laser (ECL) working at

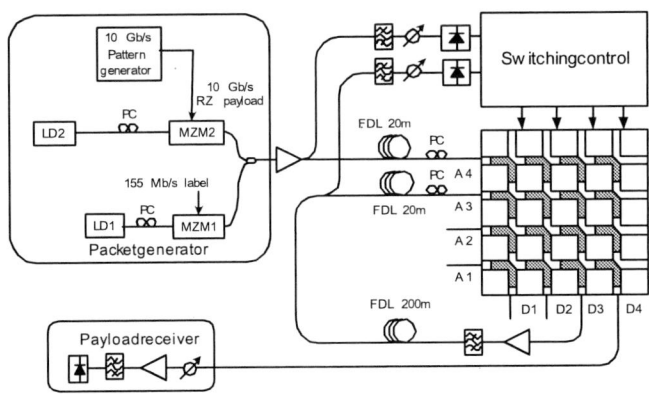

Fig.2. Experimental setup. PC: polarization controller, MZM: Mach-Zehnder modulator, FDL: fiber delay line.

1-4244-0816-4/06/$25.00 ©2006 IEEE   31

1550 nm. The 10 G/s electrical signal is generated by a programmable pattern generator. The optical payload is generated by a subsequent MZM (bandwidth 18GHz) driven by this RZ electrical data. The packets had 100ns length and were based on a 2^15-1 PRBS. The slot time in which each packet was inserted was 1.12μs. A fraction of the input packet is tapped for opto-electronic label processing. The processing takes about 32ns and the switching window is opened via the current control in less than 20ns. Therefore all the packets are buffered between the tap and the input with 20m fiber delay line for 100ns delay to ensure the switching window is fully opened when the packets arrives at the cell. Before inputting to the OXS, the payload packets are amplified, filtered, attenuated and polarization adjusted because of the polarization sensitive characteristic of OXS.

To perform the switching function, a fraction of the input packet is tapped for opto-electronic label processing. The remaining part of the packet is firstly label-erased by an optical band-pass filter and then input to the OXS. The 13 bits label signal consists of 4 bits flag at the beginning, 8 bits address and 1 bit flag in the end. The 8-bit address describes the input and output port of the OXS, which is retrieved during the routing process. This address is then mapped to the switch-cell lookup table to find out which switch cells should be in the ON-state, in order to build up an appropriate optical path along the input and output. As soon as the corresponding element of the switching cell is logic one, the control circuit will trigger the switching cell by injecting the required current. This 8-bit address can define any of the all-16 switch cells to open at any time slot.

The experiment is designed that by selecting the switch cells, the input packets will be guided from the input port into the loops, transmitted in the loops, switched from one loop to another, or guided out from the buffer. A recirculating loop with a time delay of 1.12 μs is formed by the OXS and a FDL together with inline amplification. The principle of time slot interchanging is demonstrated by the implementation of a single time slot delay. The packets to be delayed are switched into the FDL using switch A4D3. The recirculation through the buffer is handled by switch A3D3. The delayed packets are output by

switch A3D4, and the packets that are not delayed are switched by A4D4.

Fig.3 a)-e) show the 4 input packets used. In order to identify them easily, the following bit patterns were selected: Packet 1: 60ns all 1's, Packet 2: PRBS, Packet 3: 20ns 1's, 20ns 0's, and 20ns 1's (i.e. a 101 shape), Packet 4: 40ns 1's, 20ns 0's (i.e. a 110 shape). In the experiment, the order of the packets was interchanged as follows, by delaying every second packet arriving at the input to the OXS switch:

Input order: Packet1 – Packet2 – Packet3 – Packet4. Output order:Packet 2 – Packet 1 – Packet 4– Packet 3.

Fig.3 f) shows the new order of the packets leaving the TSI. The output OSNR of the signal is in the order of ~ 35dB (measured with an optical spectrum analyzer with 0.1nm optical bandwidth). The extinction ratio of the back-to-back signal is larger than 12dB, and for the switched signal is about 10dB.

Fig.4 shows the BER performance of the signal at different points of the setup. The inset figure show the waveform output of the switch and input to the buffer. The BER performance after optical switch is degraded about 1 dB. The optical buffer introduces extra degradation. The overall penalty for optical switching and buffering for 9μs is 7dB. Hence, packets arriving simultaneously to an optical node can be handled appropriately, switched or optically buffered. It is envisaged that a better performance can be obtained if the insertion loss and

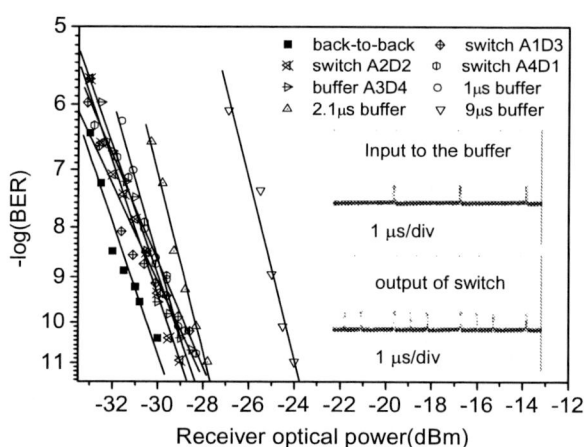

Fig.4. Measured BER curves for the back-to-back case and after optical buffer.

polarization dependent loss can be further reduced for the OXS.

IV. CONCLUSION

We have reported a novel scheme of all-optical buffering and packets time slot interchanging based on an individual switching element - a 4x4 OXS and a recirculating fiber loop. The maximum buffer depth is 9μs. The power penalty for single time slot delay is about 1dB.

REFERENCES

[1] D. Hunter, et al, J. Lightwave Technol., vol.16, no.10, pp.1725-1736, 1998
[2] E. Karasan, et al, J Select Areas Commun 16, (1998), 1081-1096.
[3] M.C. Cardakli et al., Photon Technol Lett 14, (2002), 200-202.

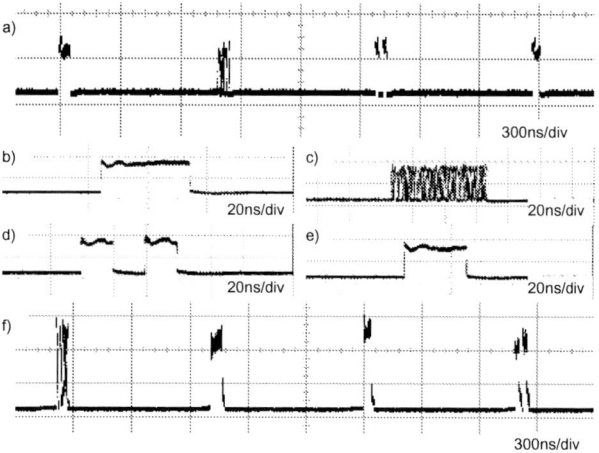

Fig.3. Measured optical waveform for optical packet buffering and time slot interchanging.

Unattended Ground Sensor System Based on Fiber Optic Disk Accelerometer

Yongjie Wang Fang Li Hao Xiao Yuliang Liu

State Key Laboratory of Integrated Optoelectronics, Institute of Semiconductors, Chinese Academy of Sciences,
Beijing 100083, PRC

Abstract—The basic principle and critical characteristics of unattended ground sensors (UGS) based on fiber optic disk accelerometers are introduced. Mechanical principles of fiber optic disk accelerometers (FODA) and calculation methods are presented. An FODA with a high sensitivity of 120rad/g and a resonance frequency of 300Hz is designed and used for detection in military affairs.

I. INTRODUCTION

Unattended ground sensors have been widely deployed in remote battlefield monitoring systems to detect, classify, and determine the direction of movement of personnel, wheeled vehicles, and tracked vehicles [1]. In traditional systems, seismic sensors are either piezoelectric, piezoresistive, or capacitive-based sensors that measure ground vibration via the current induced by the movement of men and vehicles. The response of these sensors is typically processed by a signal amplifier and converted to a voltage difference for detection and acquisition. For multiplexing and data telemetry, the traditional system requires large amounts of electronics and the associated power and telemetry cabling, which make the system expensive and heavy. Besides this, such systems are prone to fail due to electromagnetic interference (EMI) and water ingression.

With the maturation of the telecommunications industry over the past decade, the proliferation of fiber optic sensor systems has been increasing due to the improvement of several critical characteristics, including ease of multiplexing, immunity to EMI, and electrical passivity [2]. The phase modulation fiber optic sensor, a typical example based mainly on interferometric techniques, is a subject of continuing interest because of its large dynamic range and insensitivity to light fluctuations.

In this paper, we report an unattended ground sensor based on a fiber optic disk accelerometer (FODA) that employs interferometric techniques

The structure of this paper is as follows. The basic operation of the UGS system is demonstrated in section II. The construction of an all-fiber accelerometer along with its mechanical operation principles are presented in Section III. The demodulation system and experimental results are shown in Section IV.

II. DESCRIPTION OF THE UGS SYSTEM BASED ON FODA

A. Unattended Ground Sensor System

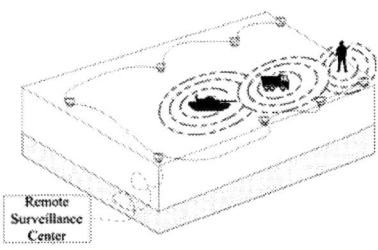

Fig. 1. UGS in military affairs.

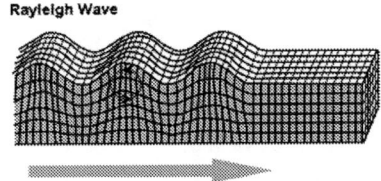

Fig. 2. Schematic diagram of a Rayleigh wave.

The basic theory of UGS is as follows: Intruders such as personnel, wheeled vehicles, and tracked vehicles initiate ground vibrations that propagate on the surface of the earth as Rayleigh waves, as illustrated in Fig. 1 and Fig. 2. The sensors are imbedded underground and detect the acceleration of the soil. We can remotely survey the area, identifying objects by pattern classification and locating their position by array application.

Several critical characteristics of this UGS based on the fiber optic disk accelerometer are as follows:

Interferometric fiber optic accelerometers provide the best sensitivity and immunity to electromagnetic interference.

There is no electrical power in the sensor head and all optical signals are confined in optical cables, ensuring that the sensors can be easily hidden without emitting any detectable radio frequency or adding any thermal signature to the environment.

A phase-generated carrier (PGC) [3] demodulation system can eliminate the influence of the initial phase of the interferometer.

Using multiplexing techniques, an array of sensors can be

1-4244-0816-4/06/$25.00 ©2006 IEEE

used to cover a large area and provide information such as direction of movement, geo-location of potential intrusions, and object identification.

B. Fiber Optic Disk Accelerometer

The fiber strain stimulated by acceleration is measured interferometrically by its effect on the phase of an optical signal. The disks and solid mass serve as the transducing medium to convert the acceleration-induced strain into an optical phase shift.

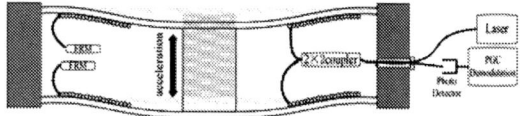

Fig. 3. Schematic diagram of an FODA.

The ground sensor head is composed of a Michelson interferometer sensor coupled with an all-fiber spring-mass configuration, as illustrated in Fig. 3. In this structure, two disks are attached to a hollow right circular cylinder. A cylindrical solid mass that rigidly connects the two disks at the inside centers insures that the plates move in unison and provides added tunable mass, which increases the accelerometer's sensitivity. Seismic waves introduce an acceleration change on the inert mass, inducing the displacement of the solid mass and subsequently resulting in strain variation on the surface of the disks. Two coils are attached to the flexural disk with glue, one on the top and one on the bottom, forming the sensing arms of a Michelson interferometer by using a Faraday rotator mirror at the end of the fiber coils. The strain on the disk surface causes tension strain in one arm and compression strain in the other. The strain in both arms of the disk is transferred to the fiber, leading to a change in the fiber length. Consequently, the phase is modulated by the acceleration.

This push-pull scheme, in which the strains on the upper disk and lower disk are equal in magnitude but 180 degrees out of phase, doubles the sensitivity and promotes noise reduction at the same time [4].

The interferometer converts this path length (phase) modulation into an intensity modulation at the 3-dB fiber coupler, which resides in the soleplate of the sensor. As a result, we can obtain the acceleration value by employing a phase-generated carrier demodulation system.

The FODA model is discussed in the next section.

III. MODEL OF FIBER OPTIC DISK ACCELEROMETER

In this section, the principles of the mechanical and optical operation of the FODA are described in the first part, and the acceleration sensitivity and frequency response are discussed in the following part.

A. Mechanical Study

In order to simplify the analysis, we make the following approximation: We ignore the mass of the disk because it is lighter than the solid mass.

The responsivity of the accelerometer can be found by calculating the displacements of different points in the disk as it flexes under acceleration and converting those displacements into the strain in the fiber [5]. In estimating the response of the sensing fiber to a vibration signal, the mechanical system of the accelerometer can be modeled by flat plates. The bending deflection of the plate is given by [6] as

$$\frac{d}{dr}\left[\frac{1}{r}\frac{d}{dr}\left(r\frac{dw}{dr}\right)\right] = \frac{Q}{D} \tag{1}$$

where Q is the shearing stress, w is the bending deflection, and $D = Et^3/12(1-\mu^2)$ is the flexure rigidity. For an edged-supported and centered-fixed plate, the edge conditions are

$$w\Big|_{r=a} = 0, \quad \frac{dw}{dr}\Big|_{r=a} = 0, \quad \frac{dw}{dr}\Big|_{r=b} = 0 \tag{2}$$

where a and b are the outer and inner radii of the disk, respectively.

From (1) and (3), we can write the bending deflection and the stress along two perpendicular directions on the surface as

$$w = \frac{Pa^2}{8\pi D}\left[(1+A)(1-\frac{r^2}{a^2}) - (B+\frac{r^2}{a^2})\ln\frac{a}{r}\right] \tag{3}$$

where $A = \frac{b^2}{a^2-b^2}\ln\frac{a}{b} - \frac{1}{2}$ and $B = \frac{2b^2}{a^2-b^2}\ln\frac{a}{b}$ depend on the geometrical parameters of the disk. The strain along the two perpendicular directions can be calculated by Hooke's law:

$$\varepsilon_r = -z\frac{d^2w}{dr^2} = \frac{3P(1-\mu^2)}{4\pi Et^2}[2(A+\ln\frac{a}{r})+1-B\frac{a^2}{r^2}] \tag{4}$$

$$\varepsilon_\theta = -\frac{z}{r}\frac{dw}{dr} = \frac{3P(1-\mu^2)}{4\pi Et^2}[2(A+\ln\frac{a}{r})-1+B\frac{a^2}{r^2}] \tag{5}$$

To calculate the change in fiber length induced by the strain on the surface of the disk, we confirm that the tangential strain is in the direction of the fiber, parallel to ε_θ. Besides this, the radial strain can "expand or reduce" the closed circle formed by the fiber. In other words, the difference in circumference is the difference in fiber length. The length change of a fiber coil is simultaneously subject to the strains along the two perpendicular directions:

$$\Delta l = \Delta l_r + \Delta l_\theta = \int_c^d (\varepsilon_r + \varepsilon_\theta)\frac{2\pi r}{\Theta}dr \tag{6}$$

When $\varepsilon_r + \varepsilon_\theta = 0$,

$$r_0 = a \cdot \exp(\frac{b^2}{a^2-b^2}\ln\frac{a}{b} - \frac{1}{2}) \tag{7}$$

From Fig. 5, we know that the direction of the total strain changes at r_0. Considering the lower bend loss of the optical signal, we conclude that we should attach the fiber coils in Area B.

B. Acceleration Sensitivity and Resonance Frequency

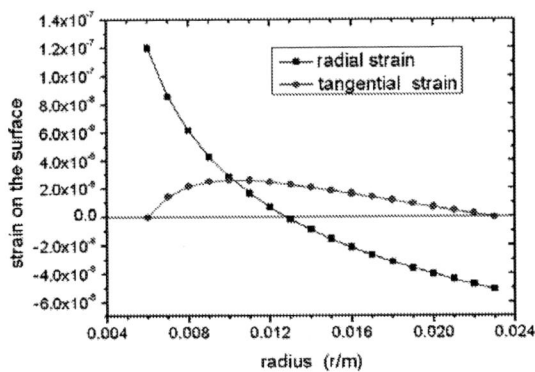

Fig. 4. Two perpendicular strains versus radius.

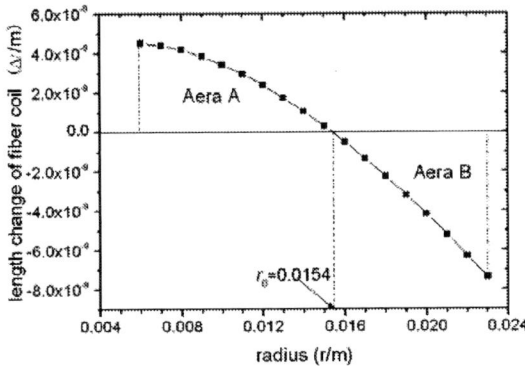

Fig. 5. Change of fiber length versus radius.

The fractional phase change depends on the changes of fiber axial length and fiber core refractive index [7]:

$$\frac{\Delta\varphi}{\varphi} = \frac{\Delta l}{l} + \frac{\Delta n}{n} \tag{8}$$

From (4), (5), (6), and (8), the normalized acceleration sensitivity can be expressed as (9)

$$\frac{\Delta\Phi}{\Delta G} = \frac{\beta}{2}\left[\left(A+\frac{1}{2}\right)\frac{d^2-c^2}{2} + \frac{1}{2}\left(d^2 \ln\frac{a}{d} - c^2 \ln\frac{a}{c}\right) \right] \tag{9}$$

where $\beta = 0.79 \times 9.8 \times 4nkmt/(D\Theta)$ is a constant associated with the geometrical parameters of this copper-based disk and the strain is directly proportional to the vibration acceleration.

The frequency response of the sensor system is related to two factors. One dominant factor is the geometric shape and material of the disk. Others are the attributes of the solid mass, aspect ratio, or stiffness in order to alter the accelerometer sensitivity and frequency range. Considering that the disk and solid mass are a mass-spring system, the maximal displacement of disk is

$$w_{max} = \frac{3P(1-\mu^2)}{4\pi Et^3(a^2-b^2)}\left[(a^2-b^2)^2 - 4a^2b^2 \ln^2\frac{b}{a} \right] \tag{10}$$

Thus the effective elastic coefficient is

$$K_{eff} = \frac{4\pi Et^3(a^2-b^2)}{3(1-\mu^2)\left[(a^2-b^2)^2 - 4a^2b^2 \ln^2\dfrac{b}{a} \right]} \tag{11}$$

The fundamental resonance frequency of the disk sensor is given by

$$f = \frac{1}{2\pi}\sqrt{\frac{K_{eff}}{m}} \tag{12}$$

Thus, by selecting the mechanical parameters of the disk and mass configuration, the natural frequency response of the system can be tuned within a certain range in adapting to the different frequencies of seismic wave sources to suit a particular application. In our system, the frequency bandwidth is designed to be 10-160 Hz.

IV. PRACTICAL SENSOR AND DEMODULATION

Fig.6 Photo of the FODA.

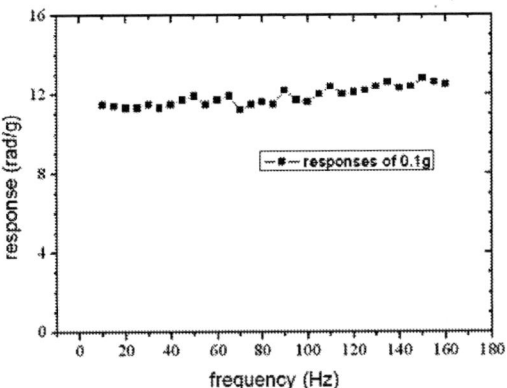

Fig. 7. Responses throughout the desired bandwidth.

In accordance with the discussion above, sensors were fabricated to test our model.

The length and the diameter of the single mode fiber are 3.43m and 250µm, respectively. The outer and inner radii of the fiber coils are 16mm and 23mm, and the outer and inner radii of the flexural disk are 6mm and 23mm. The flexural disk material is copper alloy with an elastic modulus of 1.1×10^{11}Pa. The solid mass is 17g. The acceleration sensitivity and resonance frequency are 126.64rad/g by (9), and 300Hz by (12), respectively. And Fig. 7 shows the senor sensitivity is 120rad/g in practice test.

In order to obtain a signal that does not fade because of undesired fluctuations, such as low frequency random temperature and pressure fluctuations, a phase-generated carrier homodyne method is be employed. The technique is achieved

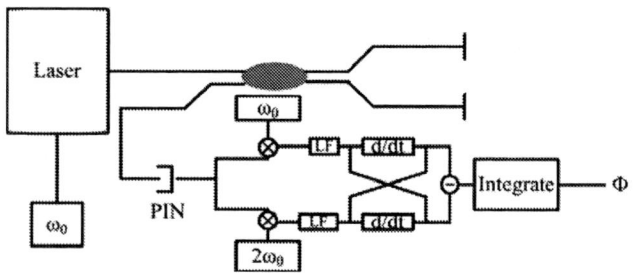

Fig. 8. Schematic diagram of PGC.

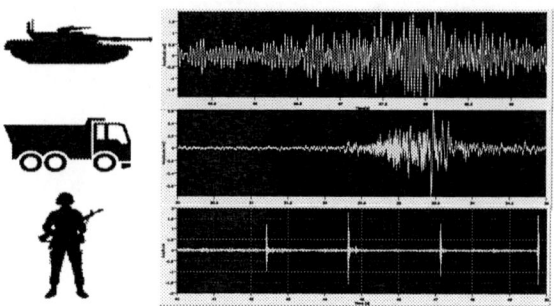

Fig.9. Acquisition in military field.

by introducing a large amplitude phase shift at a frequency outside of the signal band. Fig. 8 shows a schematic diagram of the demodulation method. A large amplitude sinusoidal modulation with a frequency outside the signal band is imposed on the drive current of the light source. The two signals received by probe detectors are multiplied by ω_0 and $2\omega_0$, respectively. Then low pass filters are used to remove the terms above the highest frequency of interest. The time derivative of these two signals, the sine and cosine terms, are cross multiplied with the cosine and sine terms, respectively, to yield the desired sine and cosine squared terms. This output can then be integrated to produce a signal that includes all of the drift information in addition to the actual signal.

The minimum measurable demodulation phase is 10^{-5}rad, and its minimum acceleration reaches 80ng.

Figure 9 shows the recorded signals from tracked vehicles, wheeled vehicles, and personnel in the military field.

V. CONCLUSION

We have discussed various aspects of UGS based on FODA. Lightweight and compact volume make it easy to carry and hide on a battlefield.

In this FODA design the spindle is attached by machine screw and nut so that it can be replaced with spindles of different mass, aspect ratio, or stiffness in order to alter the accelerometer's sensitivity and frequency range to suit a particular application.

Future work will focus on array applications by multiplexing technology, such as WDM and TDM.

Besides military battlefield this system could be used for monitoring military outposts, border protection, national landmarks, prisons, nuclear facilities and countless other sensitive areas. It has great promise for future applications.

ACKNOWLEDGMENTS

The authors would like to thank Dr. S W Zhang at the practice test for wheeled vehicles and tracked vehicles.

REFERENCES

[1] Yan Zhang, Sanguo Li, Zhifan Yin, Hong-Liang Cui, "Unattended ground sensor based on fiber Bragg grating technology," *Unattended Ground Sensor Technologies and Applications VII, SPIE*, vol. 5796, pp. 133–140, 2005

[2] Zeng N, Shi C Z, Zhang M, Wang L W, Liao Y B and Lai S R, "A 3-component fiber-optic accelerometer for well logging," *Optics Communications.* vol. 234, pp. 153–162, 2004

[3] Y Anthony Dandridge, Alan B. Teeten, Thomas G. Giallorenzi, "Homodyne demodulation scheme for fiber optic sensors using phase generated carrier," *IEEE Transactions on Microwave Theory and Techniques,* vol.30, pp. 1635–1641, 1982

[4] Yongjie Wang, Hao Xiao, Songwei Zhang, Fang Li and Yuliang Liu, "Design of a fiber-optic disk accelerometer: theory and experiment," *4th International Symposiun on Instrumentation Science and Technology,* vol. 2, pp. 995–999, 2006

[5] Geoffrey A. Cranch and Philip J. Nash,, "High-responsivity fiber-optic flexural disk accelerometers," *J. Lightwave Techn.,* vol.18, pp. 1233–1243, 2000

[6] Qinshan Fan ,"Stress analysis on axial symmetry structure," Beijing: Higher Education Press,1985,pp242

[7] C D Butter, G B Hocker, "Fiber optic strain gauge," *Appl. Opt,* vol. 17, pp. 2867–2869, 1978

40Gb/S Simultaneous Inverted and Non-inverted Wavelength Conversion Based on SOA Using Transient Cross Phase Modulation

Jianji Dong[1], Xinliang Zhang[1], Songnian Fu[2], P. Shum[2], *Senior Member, IEEE*, Dexiu Huang[1]

(1) Wuhan National Laboratory for Optoelectronics, School of Optoelectronics Science and Engineering, Huazhong University of Science and Technology, Wuhan, 430074, China.

(2) Network Technology Research Centre, Nanyang Technological University, 637553, Singapore.

Abstract—We theoretically discuss 40Gb/s semiconductor optical amplifier (SOA)-based wavelength conversion (WC) using a detuning optical bandpass filte. Both inverted and non-inverted WCs are obtained by shifting the filter central wavelength with respect to the probe wavelength when input data signal is RZ format. The filter detuning is essential to different output formations.

I. INTRODUCTION

All-optical wavelength conversion (WC) based on semiconductor optical amplifiers (SOAs) has received considerable attention during the past years in terms of small footprint, low power consumption, and optical integration [1]. However, the relatively slow gain recovery time of SOAs (typically ~100ps) limits the maximum operation speed. Currently the all-optical signal processing technique based on transient cross phase modulation (T-XPM) of SOAs is a promising technique, which not only provides high speed operation, but also has the advantage of simple configuration [2]. The structure of T-XPM consists of a SOA and a detuning optical bandpass filter (OBF). The key point is that the injected data signal should be ultrashort pulses (<10ps), and the consequent OBF has some detuning to the probe signal, that means, the OBF central wavelength is different from the probe wavelength. M. L. Nielsen presented 40Gb/s non-inverted WC by employing an OBF to select the blue-shifted part of the probe spectrum [3]. Y. Liu presented 320Gb/s inverted WC by selecting the blue-shifted probe spectrum [4].

So far, simultaneous inverted and non-inverted WCs have been reported with several techniques, such as using two cascaded SOAs based on cross gain modulation (XGM) [5], using a single-stage two-pump fiber optical parametric amplifier based on XGM and four wave mixing [6], and using single SOA followed by a detuning OBF based on T-XPM [7]. The cascaded scheme in Ref. [5] makes the system complex and the operation speed is limited due to XGM. Ref. [6] presented a simple structure but need two pump signals with high power. The T-XPM scheme in Ref. [7] is robust and promising due to simple structure and high speed operation.

In this paper, we theoretically propose both inverted and non-inverted WCs at 40Gb/s based on T-XPM technique when the input signal is RZ format. Our simulations show the detuning of the filter is essential to the output results.

II. THEORETICAL APPROACH

The WC scheme based on T-XPM can be schematically illustrated in Fig.1. A continuous wave (CW) probe beam at wavelength λ_c is combined with another modulated data signal with return to zero (RZ) format and launched into the SOA. The data signal will modulate the phase and gain of probe beam at a different wavelength trough the carrier dynamics. The subsequent OBF has some detuning to the probe signal with the central wavelength $\lambda_c + \Delta\lambda_{det}$, where $\Delta\lambda_{det}$ is the detuning value from probe wavelength λ_c. Finally, the output waveform will be displayed by the oscilloscope. The inverted WC and non-inverted WC can be obtained with different detuning of the OBF.

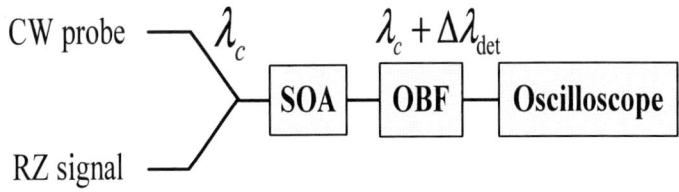

Fig. 1. The illustration of both inverted and non-inverted WCs based on T-XPM

In order to explore the intrinsic mechanism of our proposed scheme, a potent SOA model should be developed to predict the SOA operation. The theory of Agrawal and Olsson assumes that the gain saturation of the SOA is caused only by the depletion of carrier density owing to stimulated emission and neglects the main intraband processes, such as SHB and CH. However, when the SOA is operated with pulses shorter than a few picoseconds, intraband effects become important. The typical characteristic time of SHB is approximately 50-100 fs and the characteristic time of CH ranges from 700fs to 1.3ps. The theory of Mecozzi and Mørk is a simple and powerful method for calculating the amplification of picosecond pulses in a SOA [8][9]. Therefore we amend our previous model [10] with their theory associated with the intraband mechanism. Then the material gain can be expressed as $\bar{g} = g_l /(1 + \varepsilon P_{tot})$, where P_{tot} is the input total power coupled into the SOA. ε is the nonlinear gain suppression factor associated with SHB and CH effect, where $\varepsilon = \varepsilon_{CH} + \varepsilon_{SHB}$. g_l is the linear material gain coefficient, described in Ref. [10] in detail.

III. THEORECTICAL STUDY WC OF RZ FORMAT

In this section, we discuss the WC of RZ signal with ps-scale pulses by considering SOA intraband effects. When the 2.5ps-wide RZ signal is injected into a SOA with ε =0.6, the phase shift $\Delta \Phi$ and chirps $\Delta \nu$ of the probe signal are plotted in Fig. 2. We notice that, in Fig.2 (a), $\Delta \Phi$ is divided into two parts, $\Delta \Phi_N$ due to carrier density modulation and $\Delta \Phi_{CH}$ due to CH contribution, which are corresponding to the slow recovery and fast recovery of converted signal, respectively. Define that $R_{CH} = \Delta \Phi_{CH} / \Delta \Phi$, which represents the ratio of fast recovery and total recovery of converted signal. Fig.2 (b) shows the chirp variations, where the red peak and blue peak are related to the slope of the leading edge and trailing edge of the phase.

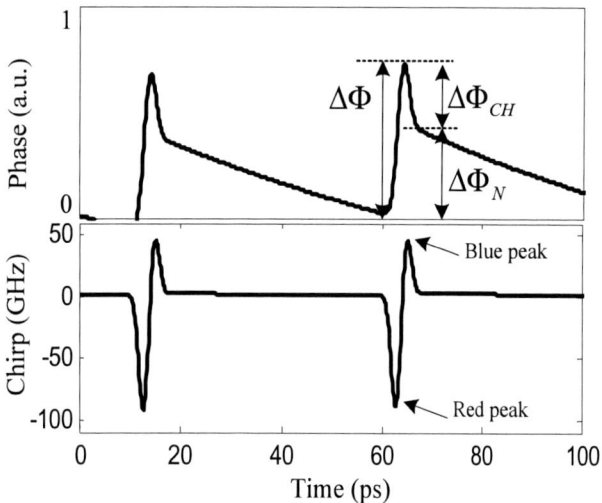

Fig. 2. The temporal phase and chirp.

Gain suppression factor ε is an important parameter, which denotes the strength of intraband effects. Fig. 3 shows the peak chirp and R_{CH} as a function of ε. Both peak chirp and R_{CH} will increase as ε increases. It interprets that the ultrafast intraband dynamic processes can help to generate large chirp and frequency shift, and to accelerate recovery of converted signal.

To investigate the impact of gain suppression factor on WC results, we fix ε =0.3 first. When the input RZ signal is 8-bit stream "11110010" with 2.5ps pulsewidth modulated at 40Gb/s, we simulate the normalized power distribution of the converted signal in time domain when the OBF detuning varies from

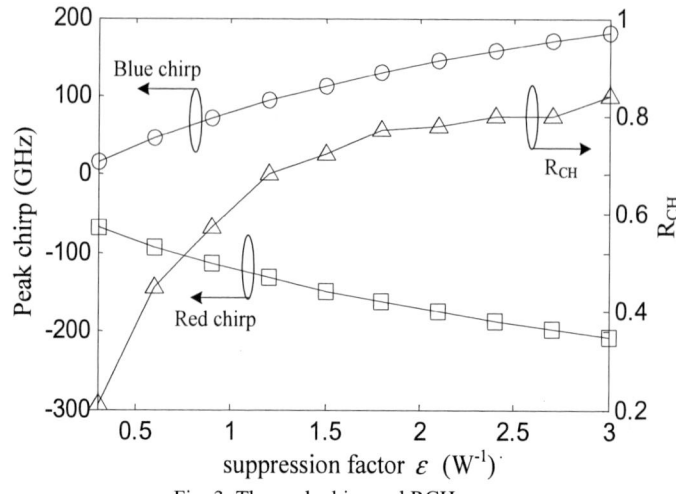

Fig. 3. The peak chirp and RCH v.s. ε

-0.8nm to 0.8nm, as shown in Fig. 4 (a). From the color figure, we find several special power distribution lines corresponding to $\Delta \lambda_{det}$ =0.8nm, -0.8nm, 0nm and -0.32nm. The crosses are the extremal points where the power is either zero or one. When $\Delta \lambda_{det}$ =0.8nm and -0.8nm, we plot the converted waveforms in Fig.4 (b). In such a case, it functions as non-inverted WC, except that some pattern effects occur. The pattern effect here is named as nonlinear patterning (NLP), which results from carrier density saturation [8]. When $\Delta \lambda_{det}$ =0nm and -0.32nm, we plot the converted waveforms in Fig.4 (c). In such a case, it functions as inverted WC, however, the converted waveform of $\Delta \lambda_{det}$ =-0.32nm has much higher extinction ratio (ER) and much weaker pattern effects than that of $\Delta \lambda_{det}$ =0nm. It reveals a blue-shifted OBF can accelerate gain recovery of inverted WC and reduce the pattern effect. The pattern effect here is defined as linear patterning, which is due to the relatively slow gain recovery and can be compensated by the blue-shifted OBF.

It is interesting that the non-inverted WC can be obtained by large OBF detuning (whether blue or red shifted), while the inverted WC can be obtained by small OBF detuning (blue shifted). This can be explained as following. The input RZ signal will induce transient nonlinear phase shifts and intensity modulation to the probe signal via cross phase modulation (XPM) and cross gain modulation (XGM) in the SOA. The nonlinear phase shifts will result in a chirped converted signal with the broadened spectrum. On the one hand, the chirped probe wavelength shift occurs only in the leading edges or the trailing edges of input RZ signal. When the OBF is detuned far away from the probe wavelength to select such probe energy at shifted components and to suppress the probe carrier, the mark of output waveforms appears in the leading edges or trailing edges of input RZ signal, and the space appears in other timeslot. As a result, the converted signal will keep in-phase to the input RZ signal. That is non-inverted WC. On the other hand, the XGM will result in the inverted WC with relatively slow recovery if no filter detuning, but the blue shifted OBF can balance the power of blue

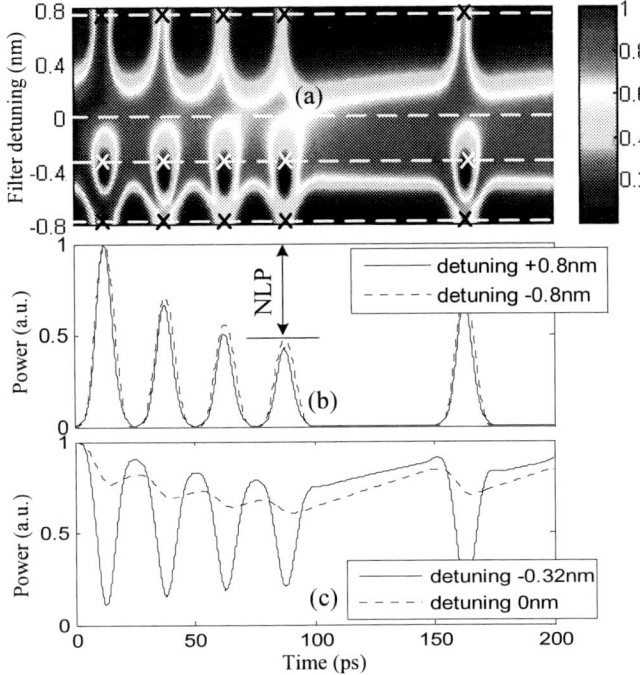

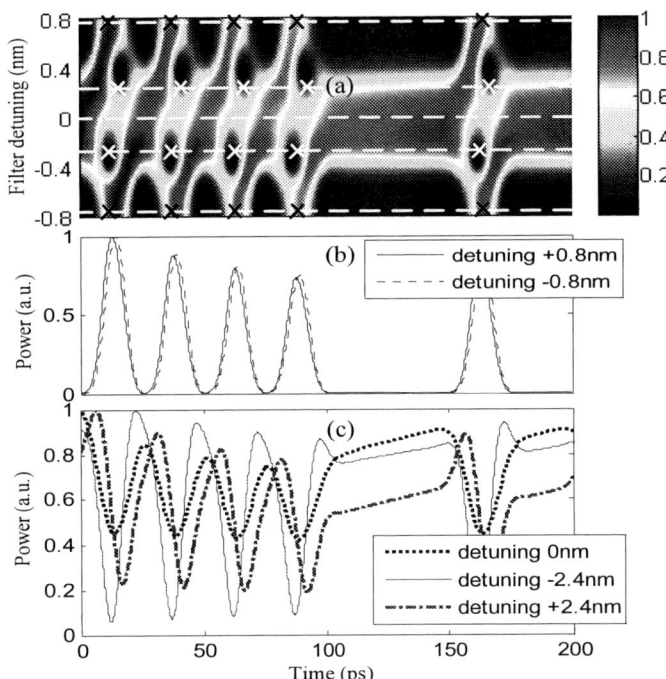

Fig. 4. The output waveforms with different detuning when ε =0.3. (a) is the normalized power distribution with different detuning. (b) is non-inverted WC when the detuning is 0.8nm and -0.8nm. (c) is inverted WC when the detuning is -0.32nm and 0nm.

Fig. 5. The output waveforms with different detuning when ε =3. (a) is the normalized power distribution with different detuning. (b) is non-inverted WC when the detuning is 0.8nm and -0.8nm. (c) is inverted WC when the detuning is -0.24nm, 0.24nm and 0nm.

chirped component and the probe power during gain recovery. The power is increasing during the gain recovery while the power of blue chirped component is decreasing. As a result, the net power at the filter output is approximately constant. This means the system effectively recovers much faster than the SOA gain. Besides, the blue shifted filter will make the gain pit much lower, therefore, the ER will be improved compared with no any detuning.

Now we reset ε =3 and simulate the converted signal power distribution in time domain when the OBF detuning varies from -0.8nm to 0.8nm, as shown in Fig. 5(a). In similar way, we find the output waveform is non-inverted WC when $\Delta\lambda_{det}$ =0.8nm and -0.8nm, however the NLP is greatly reduced, as shown in Fig. 5 (b). We find the output waveform is inverted WC when $\Delta\lambda_{det}$ =-0.24nm and the ER is greatly improved compared with no detuning, as shown in Fig. 5(c). But unlike the case of ε =0.3, it is inverted WC in the red-shifted region at $\Delta\lambda_{det}$ =0.24nm. This can be explained as following. Since the strong intraband effects can generate large chirp and accelerate recovery of converted signal, the inverted waveforms of ε =3 is much better than that of ε =0.3 even without the OBF detuning. Then a red-shifted OBF will depress the gain pit to improve the ER. This simulation can be proved by our previous experiments, which reported the non-inverted WC with large blue or red-shifted detuning and inverted WC with small blue or red-shifted detuning [7].

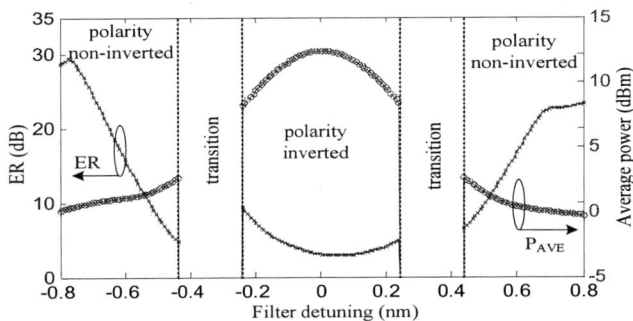

Fig. 6. The output ER and average power as a function of filter detuning.

The improved ER obtained by detuning the OBF comes at the price of reduced output power, which introduces a tradeoff between high ER and high optical signal to noise ratio (OSNR). This is illustrated in Fig. 6, which shows the output ER and average power as a function of OBF detuning. In the simulation, ε =3. As explained above, the non-inverted WC exists in both sides with large OBF detuning, while the inverted WC exists in the middle region with small OBF detuning. There are two transitional regions, which are neither inverted nor non-inverted WC.

The NLP, resulting from gain saturation, cannot be easily suppressed. Here we discuss the gain suppression factor and input puslewidth on the NLP. Fig. 7 shows the NLP variation with respect to ε when the detuning is 0.8nm and -0.8nm. The NLP will decrease as ε increases. This is so because the strong intraband effects can deepen the SOA gain modulation, then the NLP can be reduced. Fig. 8 shows the NLP variation with respect

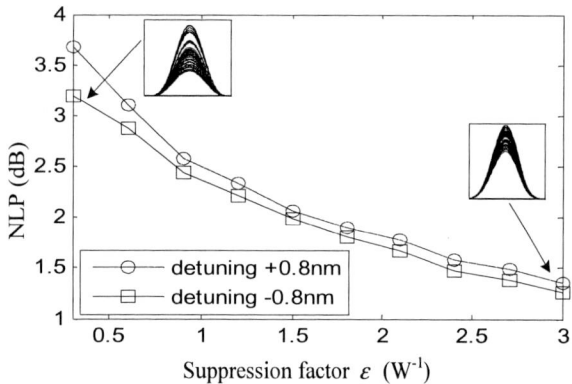

Fig. 7. NLP variation with respect to the gain suppression factor

to the RZ pulsewidth when the detuning is 0.8nm and -0.8nm, respectively. The NLP will increase as the RZ pulsewidth increases. The reason lies in that the short pulse injection will result in large phase shift of the converted signal. Then fast recovery and deep gain modulation will appear, therefore the NLP can be reduced.

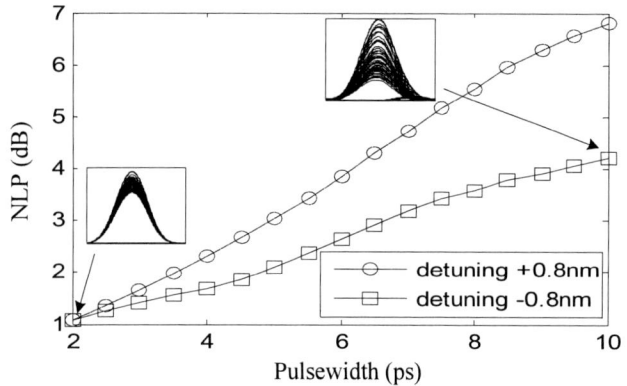

Fig. 8. NLP variation with respect to the input pulse width when the detuning is 0.8nm and -0.8nm.

IV. CONCLUSION

We have proposed and demonstrated theoretically the multifunctional WC at 40Gb/s using a single SOA followed by a detuning filter. We obtain both inverted and non-inverted WCs by different detuning of the filter.

ACKNOWLEDGMENTS

The author would like to thank Dr. Y. Yu, Dr. J. Xu, and Mr. L. Liu for the valuable discussions and assistance.This work was supported by National Natural Science Foundation of China (Grant No. 60407001), the Science Fund for Distinguished Young Scholars of Hubei Province (Grant No. 2006ABB017) and the Program for New Century Excellent Talents in Ministry of Education of China (Grant No. NCET-04-0715).

REFERENCES

[1] K. K. Chow, C. Shu, M.W.K. Mak, and H. K. Tsang, "Widely tunable wavelength converter using a double-ring fiber laser with a semiconductor optical amplifier," IEEE Photon. Technol. Lett., Vol. 14, pp. 1445-1447, 2002.

[2] S. Nakamura and K. Tajima, "Ultrafast all-optical gate switch based on frequency shift accompanied by semiconductor band-filling effect," Applied Physics Letters, Vol. 70, pp. 3498-3500, 1997

[3] M. L. Nielsen, B. Lavigne, B. Dagens, "Polarity-preserving SOA-based wavelength conversion at 40 Gbit/s using bandpass filtering," Electron. Lett., Vol.39, pp. 1334 - 1335, 2003.

[4] Y. Liu, E. Tangdiongga, Z. Li, H. deWaardt, A. M. J. Koonen, "Error-free 320 Gb/s SOA-based wavelength conversion using optical filtering," OFC 2006, PDP28, 2006

[5] A. Hamié et al., "All-optical inverted and noninverted wavelength conversion using two-cascaded semiconductor optical amplifiers," IEEE Photon. Technol. Lett., Vol. 17, pp. 1229-1231, 2005.

[6] K. K. Y. Wong, G. Lu, and L. K. Chen, "Simultaneous All-Optical Inverted and Noninverted Wavelength Conversion Using a Single-Stage Fiber-Optical Parametric Amplifier," IEEE Photon. Technol. Lett., Vol. 18, pp. 1442-1444, 2006.

[7] S. Fu, J. Dong, P. Shum, L. Zhang, X. Zhang, and D. Huang, "Experimental observations of inverted and non-inverted wavelength conversion based on transient cross phase modulation of SOA," Optics Express. Vol. 14, pp. 7587-7593, 2006.

[8] A. Mecozzi, J. Mork, "Saturation effects in nondegenerate four-wave mixing between short optical pulses in semiconductor laser amplifiers," IEEE J. Selected Topics in Quantum Electron., vol 3, pp. 1190-1207, 1997.

[9] A. Mecozzi, J. Mørk, "Saturation induced by picosecond pulses in semiconductor optical amplifiers," JOSA B, vol. 14, pp. 761-769, 1997.

[10] J. Dong, X. Zhang, Z. Jiang, and D. Huang, "Theoretical and experimental study on all-optical wavelength converters Based on the Single-port-coupled SOA," Optical and Quantum Electronics, Vol. 37, pp. 1011-1023, 2005.

40Gb/s All-optical NOR Gate Based on Semiconductor Optical Amplifier and Fiber Delay Interferometer

Jing Xu, Xinliang Zhang, Deming Liu, and Dexiu Huang
Wuhan National Laboratory for Optoelectronics and Institute of Optoelectronic Science and Engineering,
Huazhong University of Science and Technology, Wuhan, 430074, P.R.China

Abstract—**An ultrafast all-optical logic NOR gate based on a semiconductor optical amplifier (SOA) and a fiber delay interferometer (FDI) is presented. For high-speed input return-to-zero (RZ) signal, nonreturn-to-zero (NRZ) switching windows which satisfy Boolean NOR operation can be formed by properly choosing the delay time and the phase shift of FDI. 40Gb/s NOR operation has been demonstrated successfully with low control optical power. Simulations show that even higher speed can be expected.**

I. INTRODUCTION

All-optical signal processing would be a key technology in future high-speed optical communication networks in which all-optical logic operation is an elemental function. It can be widely used in all-optical demultiplexing, switching, buffering, regenerating, and computing etc. Nonlinear effects in semiconductor optical amplifier (SOA) have been widely investigated for realizing all-optical logic gates. A lot of schemes based on cross-gain modulation (XGM) have been reported, such as AND gates [1], NAND gate [2], NOR gates [3, 4], XOR gate [5] etc. Because of the slow carrier recovery process in the SOA, their output extinction ratio (ER) and pattern effects would be highly dependent on the operation bitrate. Recently, some kinds of logic gates have been successfully demonstrated at ultrafast operating speed, such as OR gate [6], XOR gates [7-10], and SOA-based switches, or the "AND" gates [11, 12]. In these schemes, interferometers based on cross-phase modulation (XPM) are employed to carry out delayed interference. Various type of interferometers have been investigated, such as Mach-Zehnder interferometers (MZIs) [7, 9], ultrafast nonlinear interferometers (UNIs)[8], delayed interferometers (DIs)[6, 10-12] etc.

In this letter, an ultrafast all-optical NOR gate based on the SOA-DI configuration, is proposed for the first time. The delayed interference is introduced by the fiber delay interferometer (FDI). By properly choosing the delay time and the phase shift of FDI, logic NOR operation can be realized at high speed with a SOA whose carrier recovery process is relatively slow. In this letter, 40Gb/s operation for input return-to-zero (RZ) signal has been demonstrated successfully with low control optical power. Output nonreturn-to-zero (NRZ)

signal is obtained. Note that the output RZ signal can be achieved by changing the continuous wave (CW) probe light into clock signal. Good consistency is observed between experiment and numerical simulation. Simulations at even higher operation speed have been taken out and a quality factor larger than 7 is found for the simulated 80Gb/s eye diagram.

II. OPERATION PRINCIPLE

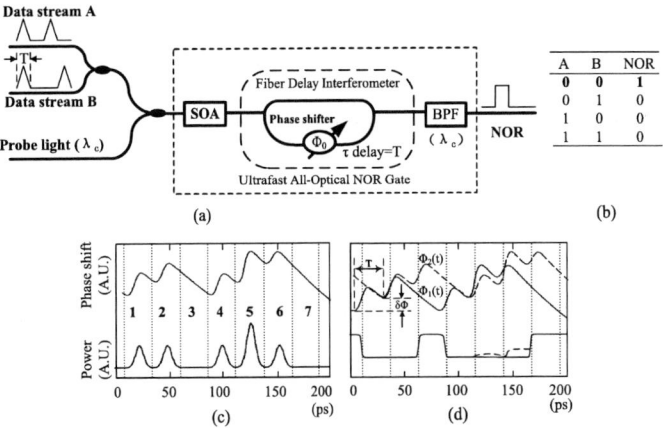

Fig. 1 Operation principle of the ultrafast all-optical NOR gate. (a) Basic configuration; (b) NOR truth table; (c) The phase variation (upper curve) modulated by the combined data stream (lower curve) in SOA; (d) Phase variations in the two arms of FDI (upper curves) and NRZ switching windows which satisfy Boolean NOR operation (lower curve).

The basic configuration of the ultrafast all-optical NOR logic gate is shown in Fig. 1(a). It consists of a SOA, a FDI, and a band pass filter (BPF). The time delay of the FDI is τ and the phase difference between two arms induced by the phase shifter is Φ_0. The BPF is used to extract the probe light and may not be necessary if the probe light is counter injected into the SOA. To illustrate the NOR operation, two input data streams are set to be "1001100" and "0100110" separately. The single bit period of these pulses is T. As shown in the lower curve of Fig. 1(c), the power combination of two data streams forms a new data of two bit "1"s. The upper curve shown in Fig. 1(c) represents the simulated phase variation of the probe light modulated by data stream "1101210" through XPM in SOA. All the simulations throughout the letter are based on a comprehensive numerical model of a traveling wave SOA [14] under the operation conditions listed in Table 1 in section 4 (special words will be given if there is any exception). As shown in Fig.1(c), during the "1" or "2" bit period (i.e. the 1st, 2nd, 4th, 5th, 6th period), phase variation of a certain profile is printed onto the probe light. Due to the fast carrier depletion and slow carrier recovery process of SOA, these profiles are quite similar to each other. Therefore, with the help of DI, the similarity of phase variation can be utilized so that the probe power in "1" or "2" bit periods can be suppressed through destructive interference. As shown in Fig. 1(d), $\Phi_1(t)$ and $\Phi_2(t)$ represent the phase variation in the upper and lower arms of FDI and satisfy the following relationship:

$$\Phi_2(t) = \Phi_1(t-\tau) + \Phi_0 \tag{1}$$

In order to suppress the probe power during the "1" or "2" bit periods, $\Phi_1(t)$ and $\Phi_2(t)$ must satisfy

$$\Phi_2(t) = \Phi_1(t-T) + \delta\Phi + (2m+1)\pi, \ m = 0,\pm1,\pm2,\cdots \tag{2}$$

Comparing Eq. (1) with Eq. (2), τ and Φ_0 of the FDI must satisfy the following relations:

$$\tau = T, \Phi_0 = \delta\Phi + (2m+1)\pi, \ m = 0,\pm1,\pm2,\cdots \tag{3}$$

where $\delta\Phi$ is physically caused by the slow carrier recovery process of the SOA and will be positive if the recovery time of SOA is relatively slower than the operation bit period. The actual value of $\delta\Phi$ varies according to the operation condition of SOA. At the same time, the probe power in "0" bit periods is preserved due to the incomplete destructive interference or even constructive interference. Therefore, if properly choose the delay time and the phase shift of FDI so that Eq. (3) is satisfied, NRZ switching windows can be formed at the output of FDI when bit "0" appears in both data streams, as shown by the lower curve in Fig. 1(d). As a result, NOR result can be obtained (see the truth table in Fig. 1(b)) and the data format of the output signal is determined by the probe light. That is, the output signal will be NRZ format if the probe light is CW signal and RZ format if clock signal is used instead.

III. EXPERIMENT SETUP AND RESULTS

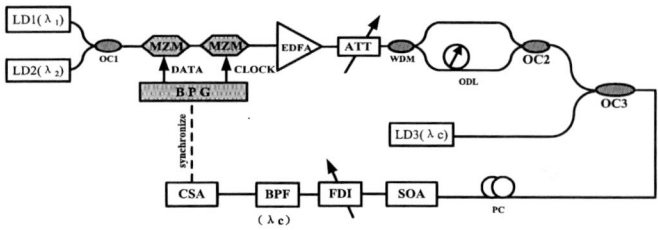

Fig. 2 Experimental Setup of the NOR gate

The experimental setup for the ultrafast all-optical NOR gate is shown in Fig. 2. In this experiment, the wavelengths of three CW signals generated by LD1, LD2, LD3 were 1549.32nm(λ_1), 1563.9nm(λ_2) and 1555.75nm(λ_C) respectively. Two cascaded

Mach-Zehnder Modulators (MZMs) were driven by data signal and clock signal respectively which were provided by the bit pattern generator (BPG). Two identical 40Gb/s data streams at different wavelength were generated at the output of the second MZM. The duty cycle of these RZ pulses was 33%. Two wavelengths were separated by the wavelength division multiplexer (WDM) and one of them was delayed one bit period by an optical delay line (ODL). Therefore, two data streams with different data pattern and wavelength were obtained at the output of the optical coulper2 (OC2). The average optical power measured at the input of SOA were -3.96dBm(λ_1), -3.69dBm(λ_2) and -2.67dBm(λ_C) respectively. The SOA is a strained multiple quantum-well device with an active region of 450μm. The driving current of SOA is 100mA. Under this current the small signal gain@1550nm is 13dB and the input saturation power is -10dBm. The 90%~10% recovery time of SOA, defined as the time needed for the gain compression to recover from 90% to 10% of the initial compression, was measured separately to be 100ps under this operation condition. The final results were analyzed through Communication Signal Analyzer (CSA). The time delay τ of FDI was fixed to 25ps for satisfying Eq. (3). The phase shift Φ_0 of the FDI was controlled by changing its working temperature. NOR result can be achieved by tuning the phase shift Φ_0 carefully. Experimental and corresponding simulation results are presented in Fig.3.

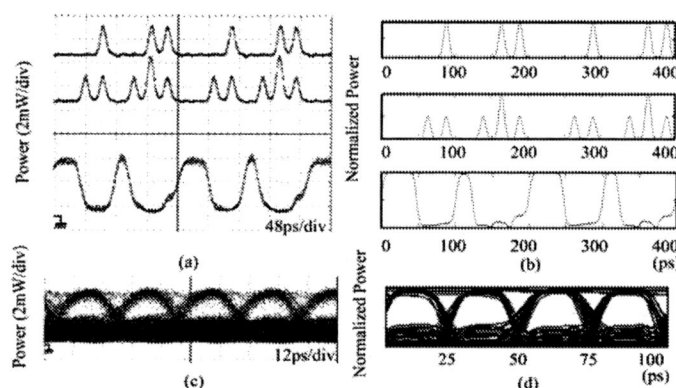

Fig. 3 40Gb/s experimental and simulated results. (a)Experimental results with fixed data pattern. Traces from the top to bottom are: data stream "00010011"; coupled two data streams of "00010011" and "00100110"; the output NOR signal in NRZ format. (c)Experimental eye diagram for 2^7-1 PRBS input data steams. (b)(d)Corresponding simulated results.

Fig. 3(a) shows the output NOR signal (bottom trace) observed by CSA with two input data streams which were fixed to be periodic "00010011" (top trace) and "00100110" respectively. The middle trace in Fig. 3(a) represents the combination of two data streams. Notice that NRZ format output signal was obtained since the probe light was CW signal. Sub-pulses are observed. The output ER was measured to be more than 10dB without considering the sub-pulses but reduced to 6dB when these pulses are taken into account. Fig. 3(c) is the eye diagram of NOR operation result for 2^7-1 pseudo random binary sequence (PRBS) data streams. An open eye is observed.

Fig. 3(b), (d) are corresponding simulated results. It should be noted that when calculating the interference, XGM were also taken into consideration for accuracy. Good consistency between simulation and experiment can be found including sub-pulses as well as the thick ground eyelid aroused by them. The sub-pulses are highly relevant to the modulated status of phase variation. The relative thin upper eyelid observed in both simulation and experiment may in a large part be the result of the following mathematical fact. The probability of two consecutive bit "1"s in NOR results is 1/16, comparing to 1/4 in original data streams where bit "1" and "0" have the same chance to appear.

Simulations show that higher operation speed can be expected. Fig. 4 shows an 80Gb/s simulated eye diagrams for $2^{}-1$ PRBS input data steams. Q factor (following the definition given in Ref. [15]) larger than 7 is obtained.

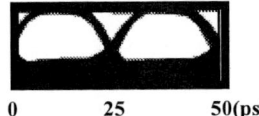

0   25   50(ps)

Fig. 4 Simulated eye diagrams for 80Gb/s $2^{'}-1$ PRBS input data steams.

IV. CONCLUSION

An ultrafast all-optical NOR gate based on the SOA and FDI is proposed for the first time. For input RZ signal, with proper delayed interference introduced by the FDI at output of SOA, NOR switching windows could be formed at the slow recovery process of SOA at high operation bitrate. 40Gb/s NOR operation has been demonstrated successfully with low optical control power. The output signal was obtained in NRZ format. However, it can be turned into RZ format by changing the CW probe light into clock signal. Simulations show that higher operation speed can be expected.

ACKNOWLEDGEMENT

The work was supported by the National Natural Science Foundation of China (Grant No. 60407001), the Science Fund for Distinguished Young Scholars of Hubei Province (Grant No. 2006ABB017) and the Program for New Century Excellent Talents in Ministry of Education of China (Grant No. NCET-04-0715).

REFERENCES

[1] X. Zhang, Y. Wang, J. Sun, D. Liu, and D. Huang, "All-optical AND gate at 10 Gbit/s based on cascaded single-port-couple SOAs," Opt. Express 12, 361-366 (2004), http://www.opticsinfobase.org/abstract.cfm?URI=oe-12-3-361.

[2] S.H. Kim, J.H. Kim, B.G. Yu, Y.T. Byun, Y.M. Jeon, S. Lee, D.H. Woo, and S.H. Kim, "All-optical NAND gate using cross-gain modulation in semiconductor optical amplifiers," Electronics Letters 41, 1027-1028(2005).

[3] A. Hamie, A. Sharaiha, M. Guegan, and B. Pucel, "All-optical logic NOR gate using two-cascaded semiconductor optical amplifiers," IEEE Photon. Technol. Lett. 14, 1439-1441(2002) .

[4] C. Zhao, X. Zhang, H. Liu, D. Liu, and D. Huang, "Tunable all-optical NOR gate at 10 Gb/s based on SOA fiber ring laser," Opt. Express 13, 2793-2798 (2005), http://www.opticsinfobase.org/abstract.cfm?URI=oe-13-8-2793.

[5] J.H. Kim, Y.M. Jhon, Y.T. Byun, S. Lee, D.H. Woo, and S. H. Kim, "All-optical XOR gate using semiconductor optical amplifiers without additional input beam," IEEE Photon. Technol. Lett. 14, 1436 - 1438(2002).

[6] Q. Wang, H. Dong, G. Zhu, H. Sun, J. Jaques, A.B. Piccirilli, and H.K. Dutta, "All-optical logic OR gate using SOA and delayed interferometer," Optics Communications 260, 81-86(2006).

[7] R. P. Webb, R. J. Manning, G. D. Maxwell, and A. J. Poustie, "40 Gbit/s all-optical XOR gate based on hybrid-integrated Mach-Zehnder interferometer," Electron. Lett. 39, 79-81(2003).

[8] R. P. Webb, R. J. Manning, and R. Giller, "All-optical 40 Gb/s logic XOR gate with dual ultrafast nonlinear interferometers," Electron. Lett. 41, 49-50(2005).

[9] Q. Wang, G. Zhu, H. Chen, J. Jaques, J. Leuthold, A. B. Piccirilli, and N. K. Dutta, "Study of all-optical XOR using Mach-Zehnder interferometer and differential scheme," IEEE J. Quantum Electron. 40, 703-710(2004).

[10] H. Sun, Q. Wang, H. Dong, Z. Chen, H.K. Dutta, J. Jaques, and A.B. Piccirilli, "All-optical logic xor gate at 80 Gb/s using SOA-MZI-DI," IEEE J. Quantum Electron. 42, 747-751(2006).

[11] Y. Ueno, S. Nakamura, and K. Tajima, "Nonlinear phase shifts induced by semiconductor optical amplifiers with control pulses at repetition frequencies in the 40-160-GHz range for use in ultrahigh-speed all-optical signal processing ," J. Opt. Soc. Am. B 19, 2573-2589 (2002).

[12] M. Nielsen and J. Mørk, "Bandwidth enhancement of SOA-based switches using optical filtering: theory and experimental verification," Opt. Express 14, 1260-1265 (2006), http://www.opticsinfobase.org/abstract.cfm?URI=oe-14-3-1260.

[13] M. L. Nielsen and J. Mørk, "Increasing the modulation bandwidth of semiconductor-optical-amplifier-based switches by using optical filtering ," J. Opt. Soc. Am. B 21, 1606-1619 (2004).

[14] Z. Jiang, X. Zhang, D. Liu, and D. Huang, "Theoretical and experimental investigation on carrier recovery time in semiconductor optical amplifier", in Semiconductor and Organic Optoelectronic Materials and Devices, Chung-En Zah, Yi Luo, Shinji Tsuji, eds., Proc. SPIE 5624, 563-574(2004).

[15] G. P. Agarwal, Fiber-Optic Communication Systems (Wiley, USA, 1997).

Sensitivity Analysis of Microring Resonator Based Biosensor: the Quality Factor Perspective

Zhixuan Xia[a] and Zhiping Zhou[a,b]

[a]Wuhan National Laboratory for Optoelectronics, Huazhong University of Science and Technology,
Wuhan, Hubei 430074, China
[b]School of Electrical and Computer Engineering, Georgia Institute of Technology,
Atlanta, Georgia 30332, USA

Abstract—The Quality factor of the dual-waveguide coupled microring resonator is investigated. By approximating the Lorenztian shape with a Guassian function, the numerical relations between the sensitivity, detection limit and the Q factor are obtained.

I. INTRODUCTION

Microring resonators are useful components for wavelength filtering [1], multiplexing [2], switching [3] and modulation [4], which are basically applied in optical communication areas. Recently, due to the relative ease of fabrication and potential high sensitivity, the microring resonator biosensors have attracted much attention and been demonstrated in various material systems [5-7]. The high sensitivity is achieved by the long lifetimes of photons that circle in the ring, which increase the probability of photons interacting with the analytes. The selectivity is dependent on the immune reactions between specified antibodies and antigens.

The Q factor is a key parameter which determines the performance of the microring biosensor, for the previous work [9] reveals that the sensitivity is largely dependent on the Q factor, moreover, the microring resonators offer a unique advantage of reducing the device dimension without sacrificing the interaction length by virtue of their high Q factor. Therefore, it needs to be closely investigated. The contents are framed as

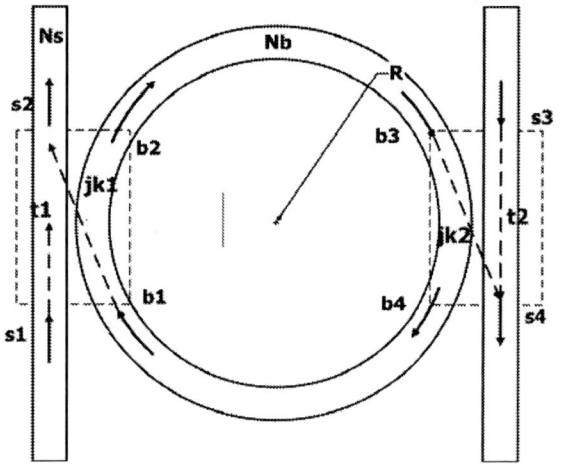

Fig.1 The schematics of a dual-waveguide coupled microring resonator. where s_i and b_i are field amplitudes to the corresponding

follows, Section II.A derives the expression of Q to the dual-waveguide · coupled microring resonator, Section II.B analyzes the numerical relations between sensitivity and the Q factor, dependence of the detection limit on the Q factor is also derived.

II. THEORY

A. The Q Factor

The Q factor has many definitions depending on various structures, among which two most popular definitions are originated from the power perspective and the spectrum perspective, respectively. The former one is defined as the ratio of the optical power stored in the cavity to the cycle averaged power radiated out of the cavity [8], while the latter one is defined as the ratio of the resonance wavelength to the full width at half maximum. Here the latter one is utilized to obtain the expression of Q factor of the dual-waveguide coupled microring resonator shown by Fig. 1.

The behavior of the microring resonator shown above is described by equations (1)-(4) as follows:

$$\begin{pmatrix} s_2 \\ b_2 \end{pmatrix} = \gamma_1 \begin{pmatrix} t_1 & jk_1 \\ jk_1 & t_1 \end{pmatrix} \begin{pmatrix} s_1 \\ b_1 \end{pmatrix} \tag{1}$$

$$\begin{pmatrix} s_4 \\ b_4 \end{pmatrix} = \gamma_2 \begin{pmatrix} t_2 & jk_2 \\ jk_2 & t_2 \end{pmatrix} \begin{pmatrix} s_3 \\ b_3 \end{pmatrix} \tag{2}$$

$$b_1 = \sigma b_4 \exp(j\varphi/2) \tag{3}$$

$$b_3 = \sigma b_2 \exp(j\varphi/2) \tag{4}$$

where σ is the transmission factor relates with the loss coefficient α by equation (5):

$$\sigma = \exp(-\alpha\pi R) \tag{5}$$

φ is the single trip phase shift which could be interpreted by effective index n_{eff}, circumference L and wave number k_0 as equation (6):

$$\varphi = k_0 n_{eff} L \tag{6}$$

Solving the above equations with the assumption $s_3=0$ leads to equation (7):

$$\frac{S_4}{S_1} = \left| \frac{s_4}{s_1} \right|^2 = \frac{\sigma^2 k_1^2 k_2^2}{\sigma^4 t_1^2 t_2^2 + 1 - 2\sigma^2 t_1 t_2 \cos(\varphi)} \tag{7}$$

Then the Q factor could be derived as the ratio of the

resonance wavelength to the full width at half maximum as expressed by equation (8):

$$Q = \frac{n_{eff}\pi L}{\lambda_m \cos^{-1}\left[2-\frac{1}{2}\left(t_1 t_2 \sigma^2 + \frac{1}{t_1 t_2 \sigma^2}\right)\right]} \quad (8)$$

Equation (8) could be rewritten by replacing the function F(v)=cos⁻¹[2-0.5×(v+1/v)] with G(v)=-In(v) (See Fig. 2) as equation (9):

$$Q \approx \frac{n_{eff}\pi L}{\lambda_0(\alpha L - Int_1 - Int_2)} \quad (9)$$

The reason of transforming equation [8] to equation [9] is to find an explicit expression of Q from which the dependence of Q on other parameters could be directly derived.

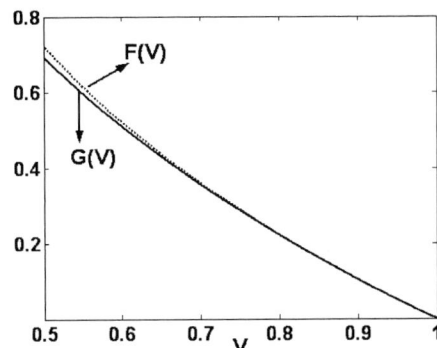

Fig.2 The plots of G(v) and F(v). Note that the values of the two functions are almost equal when v is larger than 0.7.

B. Sensitivity dependence on Q factor

As is shown by Fig. 3, the working principle of the sensor is as follows: the interaction in the sensing region causes a shift of the resonance wavelength from λ_0 to $\lambda_0+\Delta\lambda$ by changing the effective index of the propagation mode, and then the normalized output power at the fixed wavelength λ_0 will change from unity to $P(\lambda_0+\Delta\lambda)$, therefore the analytes in the sensing region could be detected by monitoring the variation of the output power.

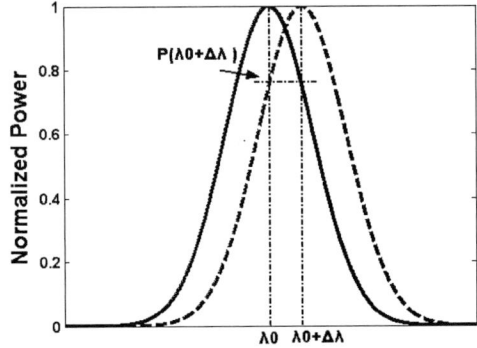

Fig. 3 Working principle of the microring resonator biosensor. The resonance wavelength varies from λ_0 to $\lambda_0+\Delta\lambda$ after the interaction occurs in the sensing region, and hence the power detected at λ_0 drops from unity to $P(\lambda_0+\Delta\lambda)$.

The sensitivity for biosensor described here is defined as the ratio of the output power to the change of the effective index, and thus it is proportional to the output power variation at a fixed working wavelength when the interaction occurs. The detection limit is defined as the minimum shift of the effective index that could be reliably detected.

Obviously, the higher sensitivity occurs when there is a larger power variation caused by a certain mount of resonance shift, and hence it is easy to understand that if the slope is sharper, the sensitivity would be higher. This is the reason why a higher Q leads to a better sensitivity. However, aside from the analytical dependence of sensitivity on Q, it is important to attain the numerical relation between the parameters, through which one could predict the detection limit of the sensor once the Q factor of the device is obtained.

The first step is to approximate the Lorentzian shape with a Guassion function, for the expression of Guassion function contains the bandwidth that is closely related to the Q factor. Then the spectrum in Fig. 3 is expressed as:

$$P_1(\lambda) = \exp\{-[(\lambda-\lambda_0)/0.6\delta\lambda]^2\} \quad (10)$$

$$P_2(\lambda) = \exp\{-[(\lambda-\lambda_0-\Delta\lambda)/0.6\delta\lambda]^2\} \quad (11)$$

Where $P_1(\lambda)$ and $P_2(\lambda)$ represent the spectrum before and after the resonance shifts $\Delta\lambda$, respectively. Note that $\delta\lambda$ is the bandwidth of the curve and thus the Q factor could be interpreted as:

$$Q = \lambda_0/\delta\lambda \quad (12)$$

Assume that the change of the effective index is Δn_{eff}, then the resonance shift is:

$$\Delta\lambda = \lambda_0\frac{\Delta n_{eff}}{n_{eff}} \quad (13)$$

Therefore, the detected power at λ_0 is calculated as:

$$P_2(\lambda_0) = \exp[-(-\Delta\lambda/0.6\delta\lambda)^2] = \exp\left[-\left(\frac{Q\frac{\Delta n_{eff}}{n_{eff}}}{0.6}\right)^2\right] \quad (14)$$

Since the sensitivity is proportional to the power variation at a fixed resonance shift, the property of $1-P_2(\lambda_0)$ reveals the numerical relations between the sensitivity and the Q factor.

In addition, the detection limit could also be obtained by equation (14). If the reliably detected normalized power change is $\delta P=1-P_2$, which is dependent on the instrument resolution and SNR, then with the given n_{eff} and Q factor, the detection limit of each device in terms of effective index could be achieved by solving equation (14) case by case.

For example, consider a silicon microing in which n_{eff} is 3.0 and assume $\delta P=10\%$, then the Q and the detection limit D has the relation as

$$Q \cdot D = 0.58 \quad (15)$$

Equation (15) means the device should have a Q factor of 10^{n-1} in order to detect an effective index change of 10^{-n}. Though the specific relation varies case by case, similar results have been demonstrated experimentally [5,9], and the agreement strongly validate our predictions.

III. CONCLUSIONS

In this paper, the Q factor of a dual-waveguide coupled microring resonator is thoroughly investigated. The explicit expression from which the Q dependence on other parameters is derived. By approximating the Lorenztian shape with a Guassian function, the numerical relations between the sensitivity, detection limit and the Q factor are obtained, which serve as guidelines for the design and optimization of the microring resonator biosensors.

REFERENCES

[1] P. P. Absil, J. V. Hryniewicz, B. E. Little, et al, "Com- pact microring notch filters," *IEEE Photonics Tech. Lett.* vol. 12, pp. 398–400, 2000.

[2] Xin Yan, Chun-Sheng Ma, Yuan-Zhe Xu, Xian-Yin Wang, Da-Ming Zhang, "Analysis and optimization for a polymer cross-grid array of microring resonant wavelength multiplexers," *Optical Engineering,* vol. 44, 075001, 2005.

[3] Vilson R. Almeida, Carios A. Barrios, Roberto R. Panepucci, Michal Lipson, "All-optical control of light on a silicon chip," *Nature*, vol. 431, pp. 1081-1083, 2004.

[4] Qianfan Xu, Bradley Schmidt, Sameer Pradhan, Michal Lipson, "Micrometre-scale silicon electro-optic modulator," *Nature*, vol. 435, pp.325-327, 2005

[5] Ketul C. Popat, John C. Aldrige, et al, "Optical Sensing of Biomolecules Using Microring Resonators," *IEEE J. Select. Topics Quantum Electronics*, vol. 12, pp. 148-154, 2004.

[6] Junpeng Guo, Michael J. Shaw, et al, "High-Q microring resonator for biochemical Sensors," *Proc. Of SPIE,* vol. 5728, pp. 83-92, 2005.

[7] C. Y. Chao, L. J. Guo, "Biochemical Sensors Based on Polymer Microrings with Sharp Asymmetrical Resonance," *Appl. Phys. Lett.*, vol. 83, pp. 1527-1529, 2003.

[8] B. E. Little, S. T. Chu, H. A. Haus, J. Foresi, and J.-P. Laine, "Microring Resonator Channel Dropping Filters," *Journal of Lightwave Tech.*, vol. 15, pp. 998-1005, 1997.

[9] Chung-Yen Chao and L. Jay Guo, "Design and Optimization of Microring Resonators in Biochemical Sensing Applications," *Journal of Lightwave Tech.*, vol. 24, pp. 1395-1402, 2006.

All-optical RZ to NRZ Format Conversion with Tunable Fiber Based Delay Interferometer

Yu Yu, Xinliang Zhang, Dexiu Huang

Wuhan National Laboratory for Optoelectronics and Institute of Optoelectronics Science and Engineering,
Huazhong University of Science and Technology, Wuhan, 430074, P.R.China

Abstract—20Gb/s all-optical format conversion from return-to-zero (RZ) to non-return-to-zero (NRZ) is demonstrated with temperature-controlled all-fiber delay interferometer (DI). The operation principle is theoretical analyzed with the help of numerical simulation and spectra analysis. Theoretical analysis results are well coincidence with experimental results. The format conversion can be achieved with low power penalty and 20dB output extinction ratio.

I. INTRODUCTION

All-optical data format conversion between return-to-zero (RZ) and nonreturn-to-zero (NRZ) may become an important interface technology for future optical networks that will include both wavelength-division-multiplexing (WDM) and optical time division multiplexing (OTDM) technologies, since RZ format is widely employed in OTDM networks as it has large tolerance to polarization mode dispersion (PMD), and NRZ format is spectrally efficient and thus is better suited for DWDM access. The conversion can be realized using nonlinear optical loop mirrors [1], ultra-fast polarization bistable vertical-cavity surface-emitting lasers (VCSELs) [2], semiconductor optical amplifiers (SOAs) [3], SOA-based interferometric devices [4],[5], or the dispersion shifted fiber (DSF) [6],[7]. However, among the proposed schemes, the SOA-based configurations, pattern effects are obvious because of the limited carrier recovery time, especially in higher operation bitrates [8]; the use of silica fiber as the nonlinear medium is more attractive compared with the semiconductor devices due to its ultrafast nonlinear response, but twisting of a long fiber or high input power are required to get enough nonlinear effect.

In this letter, we demonstrate 20Gb/s all-optical RZ-to-NRZ format conversion with all fiber delay interferometer. The passive DI converter, without exploiting any active device, has extremely simple configuration, and it is transparent to bitrates and input powers. Additional noise and pattern effect, which limit the conversion performance in other schemes with SOAs, will not appear in this proposed scheme. Simultaneously, the phase difference between two arms can be tuned continuously by tuning the operation temperature properly. We analyze this process with the help of interferential principle and spectra variation analysis (i.e. the spectra before and after conversion). The spectra analysis in frequency domain can illustrate the

conversion principle simply.

II. THEORETICAL ANALYSIS

A. Operation principle

Fig.1 shows the operation principle for format conversion based on a DI. The input signal is split into two paths along the two arms by a 50:50 coupler, one is introduced Δt (Δt <T, T is bit period) time delay, while the other is introduced a phase shift of $\Delta \varphi$. In our converter, the phase shift can be tunable by controlling the operation temperature of two arms. The two lights superimpose at another 50:50 coupler after traveling the DI. If Δt and $\Delta \varphi$ are adjusted properly, the RZ data can be

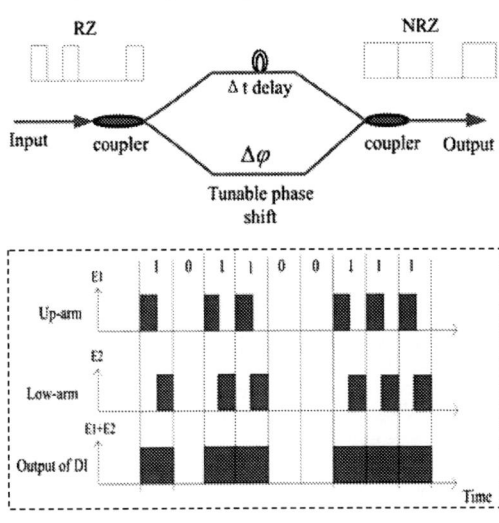

Fig.1. The principle for format conversion from RZ to NRZ based on a DI

converted to NRZ data.

Based on the interferential principle, and assuming that the coupling ratio of the coupler is 50:50, the output of DI is the function of Δt and $\Delta \varphi$, it can be expressed as

$$P_{out} = E_i^2 (\tfrac{1}{2} - \tfrac{1}{2} \cdot \cos(\tfrac{2\pi c}{\lambda} \cdot \Delta t + \Delta \varphi + \tfrac{\pi}{2})) \qquad (1)$$

Where E_i is the electric field amplitude of input RZ signal, c is the velocity of light in vacuum, λ is the center wavelength of RZ signal.

For 20-Gb/s RZ signal with a duty cycle of 50%, we fix Δt at

1-4244-0816-4/06/$25.00 ©2006 IEEE

25ps (i.e. T/2). As shown in Fig.1, if $\Delta\varphi$ is adjusted properly and constructive interference condition can be matched, the RZ signal will be converted to NRZ signal due to the constructive interference of the two signals from two arms. The duty cycle of the RZ signal is extended such that the high-power level of bit "1" is sustained over the whole bit period.

B. Transmission Spectra of DI

In order to illustrate the operation principle with spectrum analysis, the transmission spectra of DI for different conditions are analyzed.

According to Eq. (1), the DI can serve as a wavelength-selective filter (comb filter) in frequency domain. Its wavelength spacing in transmission spectrum is governed by:

$$\Delta\lambda = \frac{\lambda^2}{c \cdot \Delta t} \tag{2}$$

The maximal/minimal transmission wavelengths are determined by the phase difference of two arms. The phase difference is the integration of Δt and $\Delta\varphi$. To facilitate discussion, we assume that the phase shift induced by the time delay will be compensated by the latter one, and the phase difference of the two arms is just the phase shift $\Delta\varphi$ in one arm.

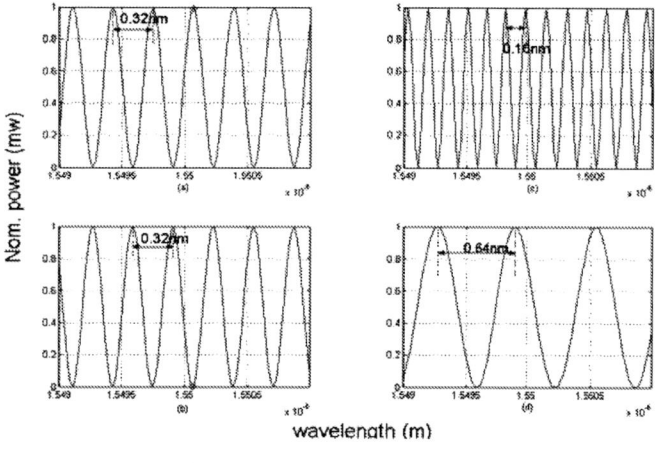

(a)　　Δt =25ps, $\Delta\varphi = 0°$　　(c)　Δt =50ps, $\Delta\varphi = 0°$

(b)　Δt =25ps, $\Delta\varphi = 180°$　　(d)　Δt =12.5ps, $\Delta\varphi = 0°$

Fig.2. Simulated transmission spectrum of DI with different conditions

Fig.2 shows the simulated transmission spectrum of DI where Fig.2(a), 2(c) and 2(d) represent Δt of 25ps, 50ps and 12.5ps, respectively. According to equation (2), the wavelength spacing is 0.32nm, 0.16nm and 0.64nm, respectively. With a fixed Δt of 25ps and different $\Delta\varphi$, Fig.2(a) and 2(b) have the same wavelength spacing but the peaks and the notches on the transmission spectrum curves are corresponding to different wavelengths. The peak wavelengths in Fig.2 (a) are corresponding to the minimal ones in Fig.2 (b), when $\Delta\varphi$ changes from $0°$ to $180°$. The maximal/minimal transmission wavelengths of DI will change with the variation of $\Delta\varphi$, which can be tunable continuously by changing the operation temperature. In other words, the transmission

spectrum of DI can be shifted without distortion by controlling the temperature.

Our proposed DI has a time delay about 25ps in one arm which is corresponding to about 5.2mm fiber length difference, and the wavelength spacing is approximate 0.32nm which is in good agreement with the calculation. This spacing decides that the DI will be suitable for 20Gb/s RZ data with 50% duty in which the interval of its two sidebands is 0.32 nm. In order to match with the signal wavelength, one of the maximal wavelengths was adjusted to aim at 1550.01nm by controlling $\Delta\varphi$. Using the ASE spectrum of EDFA, the measured transmission spectrum is shown in Fig.3. The Peak-to-notch ratio is about 15.8dB, which is determined by the coupling ratio and the loss difference of the two arms. If the coupling ratio was not just the ideal case (50:50) or the constructive interference happened non-ideally, there would be ripples on the top of the converted NRZ signals, with the degradation of extinction ratio (ER) simultaneity. In our experiment, the coupling ratio is close to 0.5, and the loss difference in two arms is very small and can be ignored.

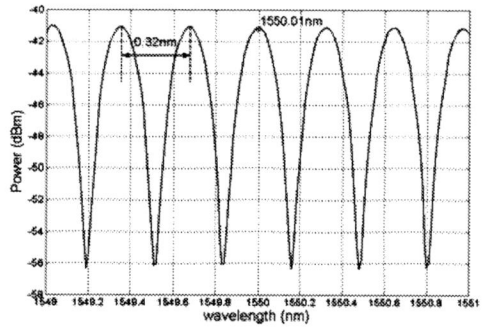

Fig.3. Measured transmission spectrum of DI with $\Delta t \approx 25\,ps$

III. RESULTS AND DISCUSSION

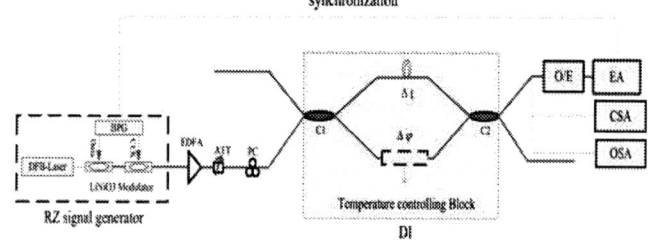

Fig.4. Experimental setup for RZ to NRZ conversion using a DI

The experimental setup for format conversion is shown in Fig.4. The SHF 40-Gb/s optical communication system is used to generate input RZ signal with wavelength of 1550.01nm, and an average power of 0dBm. Its bit rate is fixed at 20Gb/s in our experiment. The signal power can be controlled by followed erbium doped fiber amplifier (EDFA) and the attenuator (ATT). One arm is temperature-controlled to change the refractive index, thus the phase difference ($\Delta\varphi$) between the two arms will change, and different output results can be analyzed by the wavelength of the RZ signal, while two notches aim at the two

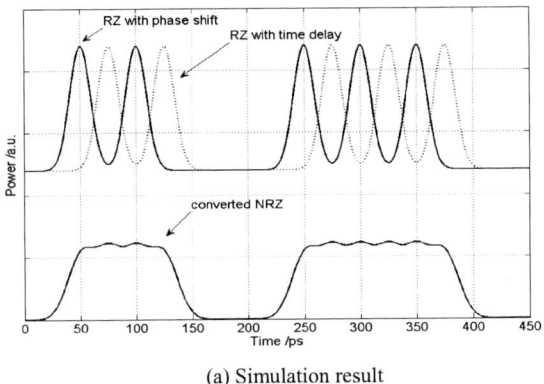

(a) Simulation result

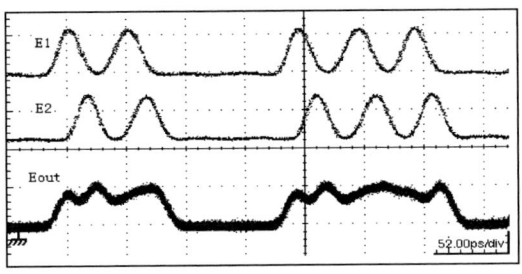

(b) Experimental result

Fig.5. RZ-to-NRZ format conversion of bit stream "11001110"

Optical Spectrum Analyzer (OSA) and Communication Signal Analyzer (CSA) respectively. When $\Delta\varphi$ and Δt are appropriate, the RZ to NRZ conversion can be achieved. The BER can be measured by the Error Analyzer (EA).

The conversion of bit stream "11001110" at 20Gb/s is shown in Fig.5. Fig.5 (a) is the simulation result, while Fig.5 (b) is the corresponding experimental result. The results reveal that the duty cycle of the input RZ signal with duty cycle of 50% is extended such that the high-power level of bit "1" is sustained over the whole bit period, and the RZ format can be converted to NRZ format due to the 25ps delay and the constructive interference. Simulation result is well coincident with experimental result.

The eye diagrams of the input RZ signal and the output NRZ signal at 20-Gb/s are presented in Fig.6. The full-width at half-maximum (FWHM) of input RZ signal is about 25ps. After passing through the DI, the converted output NRZ signals show clear and open eyes, with little ripple on the top of converted NRZ. No additional noise and pattern effect can be found. The output ER is over 20dB. BER measurement results for original RZ signal and converted NRZ signal are shown in Fig.7. Result shows that power penalty induced by the pattern converter is about 0.54dB. Considering the sensitivity difference between RZ signal and NRZ signal of BER test system, the power penalty is very small.

In order to get a comprehensive understanding of the conversion, the spectra of the input RZ signal, the converted NRZ signal together with the optimal transmission spectrum of the DI are shown and analyzed in Fig.8. From these spectra, we can see that one of the peak wavelengths of DI aims at the center

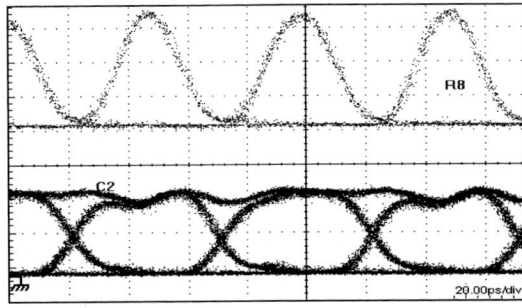

Fig.6. The eye diagrams (2^{31}-1) of input RZ (upper) and converted NRZ (lower) at 20Gb/s

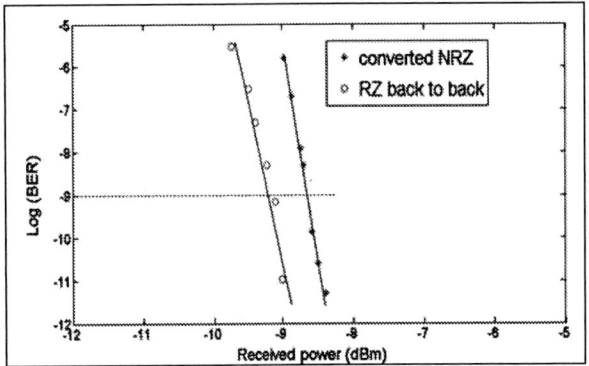

Fig.7. BER measurements for original RZ signal and converted NRZ signal

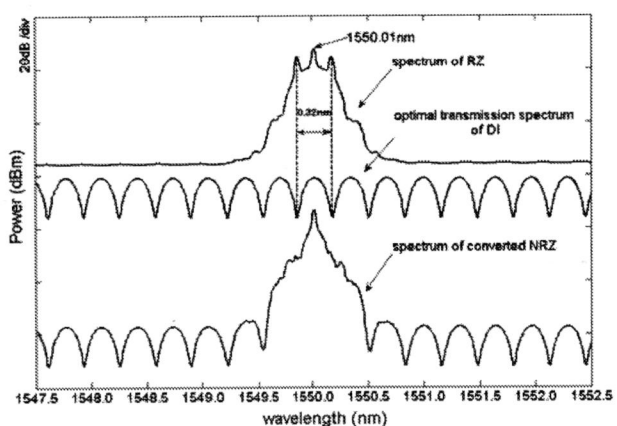

Fig.8. Measured spectrum of the input RZ (upper), the transmission spectrum of DI in our experimental condition (middle) and the spectrum of converted NRZ (lower)

sidebands, whose spacing is just 0.32nm. The carrier wavelength of the RZ can pass though the comb filter, but the two sidebands of the RZ are suppressed due to the transmission characteristic of the DI, with the reservation of the data modulation. So a NRZ-like spectrum can be achieved.

In all the discussion above, the duty cycle of RZ signal is 50%. For RZ signal with larger duty cycles, our converter is competent apparently; and for RZ signal with smaller duty cycles, which have stronger spikes in their spectra, we can change the time delay Δt or cascade two stages of our proposed DI to suppress the residual spikes deeply. Then the converter can be applied for RZ signals with other duty cycles.

In addition, if the time delay (Δt) can be adjusted to other values, such as 12.5ps or 6.25ps, naming the wavelength spacing of the comb filter is 0.64nm (40-GHZ) or 1.28nm (80-GHZ), the RZ to NRZ format conversion can be achieved at bit rate of 40-Gb/s or 80-Gb/s. Furthermore, other format conversion can be achieved using the same configuration, if the phase shift $\Delta\varphi$ is fixed at other value. The operation temperature of the DI can be tuned continuously thus its transmission spectrum would be shifted without distortion, so our converter can be operated at other wavelengths. Furthermore, due to the all-passive devices, the converter is polarization and power insensitive.

IV. Conclusion

In conclusion, we have demonstrated all optical RZ to NRZ format conversion at 20Gb/s by utilizing all fiber delay interferometer. The operation principle is analyzed with the help of numerical simulation and spectra analysis. Clear and open eyes show good output performance of the converter, and small power penalty is achieved. The potential of format conversion with other duty cycle is discussed. In addition, higher bit rate and other applications can be achieved by adjusting the parameter of the DI.

Acknowledgment

Related project is supported by New century excellent talent project in Ministry of Education of China (Grant No. NCET-04-0715).

References

[1] S. Bigo, O. Leclerc, and E. Desurvire, *IEEE J. Select. Topics Quantum Electron.*, **3**, 1208–1222 (1997).

[2] H. Kawaguchi, Y. Yamayoshi, and K. Tamura, in Lasers and Electro-Optics Conference, 2000 OSA Technical Digest Series (Optical Society of America, Washington, D.C., 2000), pp. 379–380

[3] D. Norte and A. E. Willner, *IEEE Photon. Technol. Lett.*, **8**, 712–714 (1996).

[4] Chung Ghiu Lee, Yun Jong Kim, Chul Soo Park, Hyuek Jae Lee and Chang-Soo Park, *J. Lightw. Technol.*, **23**, 834–841 (2005).

[5] Lei Xu, Bing C. Wang, Varghese Baby, Ivan Glesk, *IEEE Photon. Technol. Lett.*, **15**, 308-310 (2003).

[6] C. H. Kwok, and Chinlon Lin, *IEEE J. Sel. Topics Quantum Electron.* *12, 451-458 (2006)*

[7] S. H. Lee, K. K. Chow and C. Shu, *Opt. Exp.*, **13**, 1710-1715 (2005)

[8] Qianfan Xu, Minyu Yao, Yi Dong, Wenshan Cai and Jianfeng Zhang, *IEEE Photon. Technol. Lett.*, **13**, 1325–1327 (2001).

Study of Coupled-Resonator-Induced Transparency in 3×3 Coupler Based Dual Microring Resonators

Xiaobei Zhang, Dexiu Huang and Xinliang Zhang

Wuhan National Laboratory for Optoelectronics and School of Optoelectronics Science and Engineering,
Huazhong University of Science and Technology, Wuhan, 430074, Hubei, People's Republic of China

Abstract—A novel realization of coupled-resonator-induced transparency is obtained in 3×3 coupler based dual microring resonators. The normalized intensity transmissions are discussed under different conditions using the transfer matrix model. Moreover, we adopt the finite-difference-time-domain method to numerically simulate the intensity transmissions. Simulations are found in agreement with the results obtained by using the transfer matrix model.

I. INTRODUCTION

Electromagnetically induced transparency (EIT) [1] in atoms is an interesting research topic in physics both in theory and experiments, where optical transparency is realized in multi-level mediums at a resonant transition induced by application of a coherent electromagnetic field at an adjacent transition. Similar to EIT, coupled-resonator-induced transparency (CRIT) [2-6] caused by interferences in multiple-resonators is also a charming phenomenon, which may be of great importance for optical delay lines, buffers and slow light. The experimental observation of CRIT in coupled microspheres due to interferences between their respective coresonant whispering-gallery modes (WGMs) is demonstrated [5]. In [6], the realization of CRIT spectrum is presented in integrated micron-size silicon optical resonator systems. We have proposed and analyzed four different types of dual microring resonators incorporating the 3×3 coupler with a universal model developed by using the transfer matrix method [7].

This paper describes CRIT in dual microring resonators incorporating a 3×3 coupler, which is more compact than those [2, 4, 6] due to the elimination of two parallel straight waveguides and more controllable than those [3, 5] due to the presence of one straight waveguide between dual microring resonators. With recent advances in planar fabrication techniques for optical waveguides, the device can be easily realized by laterally coupled or vertically coupled schemes and also has potential applications in optical filters, delay lines and sensors.

II. STUDY OF CRIT IN 3×3 COUPLER BASED DUAL MIRCORING RESONATORS

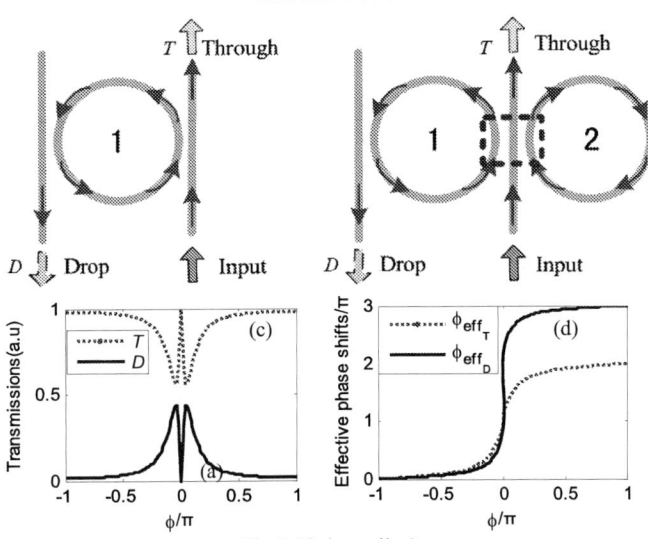

Fig.1. (Color online)
(a) Illustration of a conventional microring resonator
(b) Illustration of dual microring resonators incorporating the 3×3 coupler
(c) normalized intensity transmissions under no losses
(d) effective phase shifts under no losses

The conventional microring resonator is shown in figure 1(a), which can be generally decomposed into two 2×2 couplers and a ring resonator. What happens when an additional microring resonator is posited closed to one of the 2×2 couplers? It is obvious that one of the 2×2 couplers will turn into a 3×3 coupler.

Figure 1(b) illustrates the proposed dual microring resonators, with the dashed box corresponding to the 3×3 coupler. Compared to series-coupled dual microring resonators, a straight waveguide inserts into two resonators, which makes the coupling between resonators more complicated. Due to the introduction of the 3×3 coupler, light will couple to the two ring resonators. Details of the transfer matrix model can be found in [7]. T and D represent normalized intensity transmissions at the through and the drop ports respectively. As shown in figure 1(c), there is a narrow transmission peak between two traps corresponding to CRIT at the through port (dotted line). The spectrum of D (solid line) is complementary. In series-coupled dual microring resonators with small loss [3, 5], the classical destructive interference gives rise to CRIT. Moreover, CRIT is caused by the interference between direct and indirect pathways for two cavities' decays in parallel-coupled dual microring resonators [2, 4, 6]. However, CRIT here is caused by the introduction of the 3×3 coupler to make the destructive interference. As shown in figure 1(d), ϕ_{eff_T} (dotted line) and ϕ_{eff_D} (solid line) are effective phase shifts at the through and the drop ports respectively, and group delays are given by $T_\xi = d\phi_{eff_\xi}/d\omega$ with $\xi = T, D$. ϕ_{eff_T} possesses accessional π phase shift and its slope is much more steeper compared to ϕ_{eff_D}.

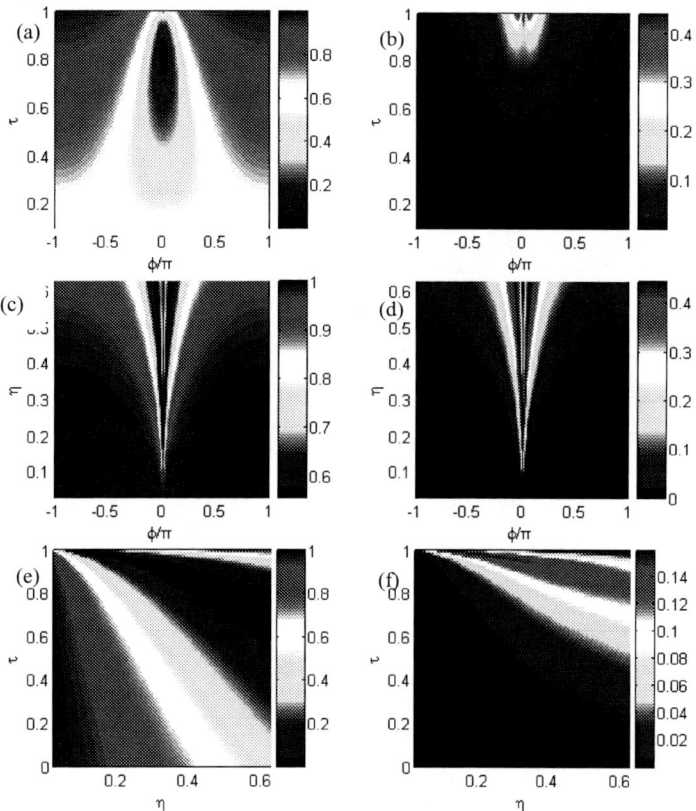

Fig.2. (Color online)
Distributions of normalized intensity transmissions under different conditions
(a) $\eta = 0.5$, T with different τ and ϕ
(b) $\eta = 0.5$, D with different τ and ϕ
(c) $\tau = 1$, T with different η and ϕ
(d) $\tau = 1$, D with different η and ϕ
(e) $T(\phi = 0)$ with different τ and η
(f) $D(\phi = 0)$ with different τ and η

When possible losses caused by absorption, scattering or radiation and different coupling coefficients are taken into account, figure 2 gives the distributions of normalized intensity transmissions under different round trip transmission coefficients τ, coupling coefficients η and normalized phases

ϕ. As shown in figure 2(a), $T(\phi = 0) = 1$ comes into existence if and only if $\tau = 1$. CRIT deteriorates as τ decreases, with the peak and two neighboring traps degenerating into just one trap. There is a region corresponding to critical coupling $T = 0$ when τ is about 0.815. As τ decreases, D also deteriorates as shown in figure 1(b), and a dip between two peaks remarkably preserves when τ is greater than 0.95. Figure 2(c) and (d) are T and D respectively with different η and ϕ. As η increases, both the distance between two traps and the 3dB width of the narrow transmission peak increases as shown in figure 2(c). Energy conservation ($T + D = 1$) comes into existence when losses are neglected (i.e. $\tau = 1$), validating the accuracy of the model. Figure 2(e) and (f) are $T(\phi = 0)$ and $D(\phi = 0)$ respectively with different τ and η. There are two separate regions where $T(\phi = 0)$ is locally maximal as shown in figure 2(e). The maximal and minimal transmission regions can be easily obtained from these distributions, which may be useful for practical designs.

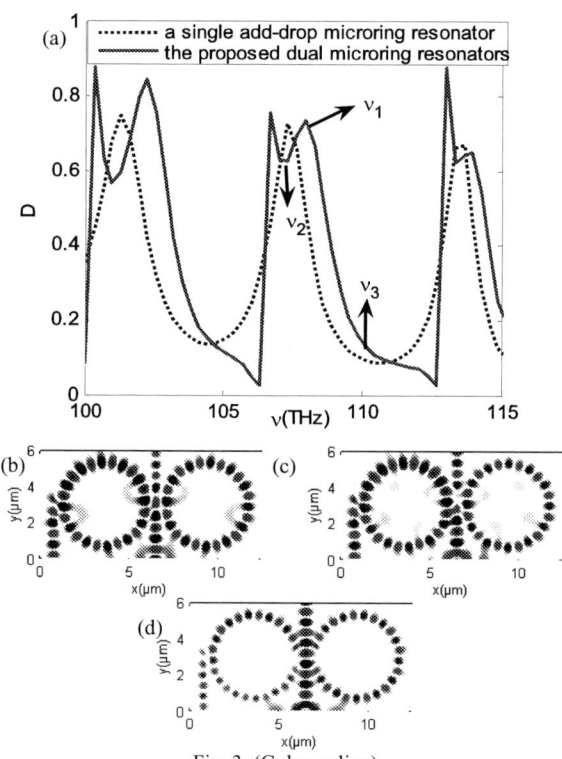

Fig. 3. (Color online)
(a) FDTD-computed transmissions at the drop ports
(b) Steady-state electric field pattern with $\nu_1 = 107.94$THz
(c) Steady-state electric field pattern with $\nu_2 = 107.20$THz
(d) Steady-state electric field pattern with $\nu_3 = 110.00$THz

The finite-difference-time-domain (FDTD) method is adopted to validate the reliability of the transfer matrix model [7]. Corresponding parameters of the structure and the materials are chose the same as [8]. By launching a Gauss pulse into the resonators, the transmission spectrum (solid line) at the drop port is calculated as shown in figure 3(a) with a single add-drop microring resonator's transmission spectrum (dotted line) also calculated for comparisons. It is obvious that there is a trap between two peaks and the trap locates nearly at the peak of the single add-drop microring resonator's transmission spectrum. Despite the high radiation loss in numerical simulation, the spectrum is still in agreement with that as shown in figure 1(c) on the whole, which suggests the presence of CRIT. As the frequency increases, the distance between two peaks becomes closer and the 3dB width of the single add-drop microring resonator decreases. $v_1 = 107.94\text{THz}$, $v_2 = 107.20\text{THz}$ and $v_3 = 110.00\text{THz}$ are chose to correspond to the peak, the trap and the low transmission point respectively, and steady-state electric field patterns are computed as shown in figure 3(b), (c) and (d). Due to the small contrast between $D(v_1)$ and $D(v_2)$, the steady-state electric field patterns are not remarkably discriminative. Because $D(v_3)$ has a much smaller value than both $D(v_1)$ and $D(v_2)$, the steady-state electric field pattern evidently differs from those. As shown in figure 3(d), there is a high transmission at the though port, a low transmission at the drop port and low intensities in the dual microring resonators.

III. Conclusion

Coupled-resonator-induced transparency in dual microring resonators incorporating the 3×3 coupler has been demonstrated. Dependences of normalized intensity transmissions on round trip transmissions, coupling coefficients and normalized phases are discussed. Finally, FDTD simulations are adopted to calculate the transmission spectrum, which is in agreement with the results of the transfer matrix model in principle. The proposed novel realization of CRIT has potential applications in optical filters, delay lines and sensors.

ACKNOWLEDGMENT

This work was sponsored by the National Natural Science Foundation of China (Grant No. 60577007) and and the Program for New Century Excellent Talents in University, Ministry of Education of China (Grant No. NCET-04-0715).

References

[1] S.E. Harris, "Electromagnetically Induced Transparency," Phys. Today, 50,No.7,36-42(1997).

[2] S. T. Chu, B. E. Little, W. Pan,T. Kaneko, and Y. Kokubun, "Second-order filter response from parallel coupled glass microring resonators," IEEE Photon. Technol. Lett. 11, 1426-1428(1999).

[3] D.D. Smith,H. Chang,K. A. Fuller,A. T. Rosenberger and R. W. Boyd, "Coupled-resonator-induced transparency,"Phys. Rev. A 69, 063804(2004).

[4] L. Maleki, A. B. Matsko, A. A. Savchenkov, and V. S. Ilchenko, "Tunable delay line with interacting whispering-gallery-mode resonators,"Opt. Lett. 29, 626-628(2004).

[5] A. Naweed, G. Farca, S. I. Shopova, and A. T. Rosenberger, "Induced transparency and absorption in coupled whispering-gallery microresonators," Phys. Rev. A 71,043804(2005).

[6] Q. Xu, S. Sandhu, M. L. Povinelli, J. Shakya, S. Fan,and M. Lipson1, "Experimental Realization of an On-Chip All-Optical Analogue to Electromagnetically Induced Transparency,"Phys. Rev. Lett. 96, 123901(2006).

[7] Xiaobei Zhang, Wei Hong, Xinliang Zhang and Dexiu Huang, "3×3 coupler based dual microring resonators: a proposal,analysis and application," unpublished

[8] S. C. Hagness,D. Rafizadeh,S. T. Ho, and A. Taflove, "FDTD Microcavity Simulations: Design and Experimental Realization of Waveguide-Coupled Single-Mode Ring and Whispering-Gallery-Mode Disk Resonators," J. Lightwave Technol. 15, 2154-2165 (1997).

Photonic Crystal Taper for Efficient Coupling and Smooth Mode Profile Conversion

Jing Liu[a], Dingshan Gao[a] and Zhiping Zhou[a,b]

[a]Wuhan National Laboratory for Optoelectronics, Huazhong University of Science and Technology,
Wuhan, Hubei 430074, China

[b]School of Electrical and Computer Engineering, Georgia Institute of Technology,
Atlanta, Georgia 30332, USA

Abstract—The designment of Si-based 2D photonic crystal taper is presented. The software Rsoft incorporating finite-difference time-domain (FDTD) technique was used to simulate light propagation through this new structure. Plenty of methods have been tried to improve transmission coefficient and solve the problem of mode mismatch in photonic crystal devices. A taper with high transmission coefficient has been designed and optimized.

I. INTRODUCTION

The implementation of photonic integrated circuits requires the size of the photonic devices could be smaller and integrated to form all kinds of functionally powerful system on chip or on board. However, ordinary optical devices manufactured by conventional technology hardly achieve such small size. Photonic crystal (PC), which can confine and guiding lightwave perfectly even in subwavelength scale, has been the most important candidate for ultracompact photonic integrated circuits. The section sizes of PC device are normally ten times smaller than those of traditional waveguides or fiber, which would bring big mode transform loss between them. Therefore, photonic crystal taper which couple light efficiently from traditional dielectric waveguides and fibers into and out of PC devices is one of the key devices in photonic integrated circuits.

Photonic crystal tapers can change modal properties, for instance, mode size and modal number, which have great applications in different PC devices. One typical application is to couple light from fiber to photonic crystal waveguide which can cause big loss through direct coupling, the taper is used to change the mode field of light propagating from fiber to PC waveguide, which reduce the coupling loss a lot. Since transformation of modal number is also available by this taper, it can be used in the sharp angle waveguide structure to change multimode to single mode when light propagates through the corner to avoid mode leakage [1].

II. TWO TYPICAL DISIGNMENTS OF PC TAPER AND THEIR APPLICATIONS

One of the main reasons why it is hard to achieve high coupling efficiency between dielectric waveguide and PC waveguide is their intrinsical differences in guiding mechanism

and field profile [2]. The waveguide mode in dielectric waveguide is entirely forward-propagation mode while that of PC waveguide consists of both forward and backward propagation mode. Moreover, mode propagates in higher refractive index region in dielectric waveguide, but propagates in lower refractive index region in PC waveguide. Therefore,

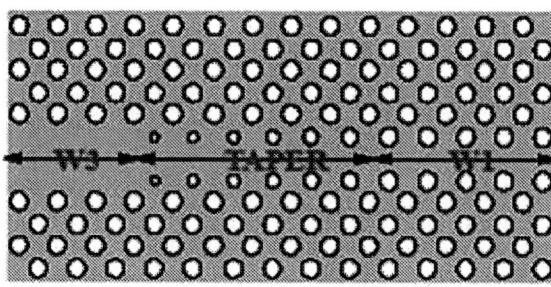

Fig. 1 the first structure

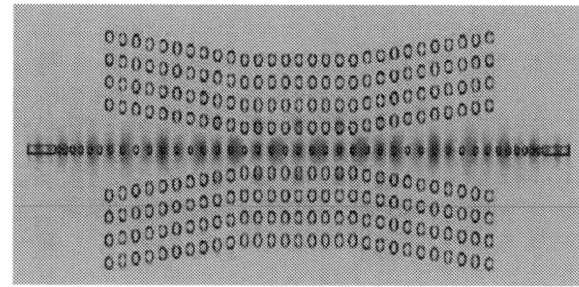

Fig. 2 the second structure

our taper is aimed to solve these two issues.

There are, at present, two typical structures as shown in Fig. 1 and Fig. 2. The principle of operation of the first taper relies on the manufacture of holes with progressively varying diameter [3]. With this variation, the central waveguide region may be seen as an artificial material with a gradient effective index. Such taper is designed on a two-dimensional(2D) PC consisting of a triangular array of holes drilled on a Si confining layer. The input and output waveguides are denoted by Wn, *which means* the PC waveguide is formed by removing n rows of air cylinders in the G*K* direction of the PC (horizontal direction in Fig. 1). Since there is large mode mismatch between a ridge waveguide and a strongly confined W1 PC waveguide, the ridge-access guide was first coupled to a W3 waveguide and then converted smoothly to the W1 waveguide through the PC taper. However, this structure only solves the mode mismatch caused by index difference, and the mode profile variation is not smooth enough when propagating through the structure.

The second structure [2], which is designed on a 2D PC consisting of a square array of Si cylinder. Actually it has two functions. One is to implement conversion between forward propagation mode in the outside waveguide and both forward and backward propagation mode in the PC waveguide by gradually decreasing the distance between the cylinders which formed a line defect. The other one is to transform the mode from high-index guiding to low-index guiding by decreasing gradually the distance between the adjacent holes and the central line defect.

Both issues discussed previously have been taken into account in the second structure, therefore, the coupling efficiency is higher than that of the first structure. However, since present PC manufacture technologies are mainly focused on drilling holes on dielectric materials, this structure can not be fabricated easily.

III. THE PROPOSED PC TAPER STRUCTURE WITH HIGH TRANSMITTANCE

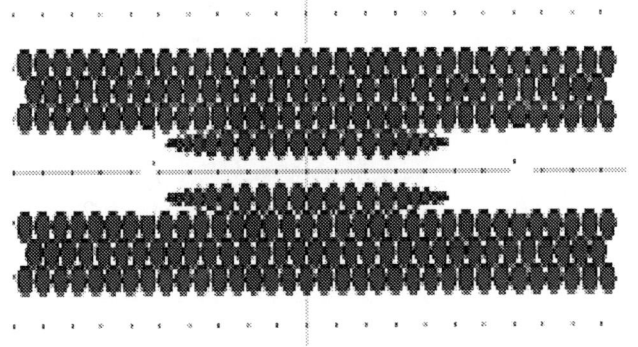

Fig. 3 the proposed PC taper structure

Therefore, we must find a new structure (as shown in Fig. 3) that combines the advantages of both structures described above, in other words, the new structure is designed on a 2D PC consisting of a triangular array of holes drilled on a Si confining layer, with the diameters and distances of holes which confine the line defect decreasing gradually. Thus, the index difference problem can be solved by the gradually decreasing of diameters, while the forward mode can be changed into both forward and backward mode by decreasing the distances between holes gradually at the same time.

The RSOFT software which is based on finite-difference time-domain (FDTD) technique was used to simulate the light propagation through the new structure with parameters of period=0.6micron, fill factor=0.62 and index=2.75. The simulation results in Fig.4(a) and Fig.4(b) show that the new structure achieves a high transmission efficiency of 85% which is higher than the first structure but lower than the second one, and a smoother mode profile transformation.

IV. CONCLUSION

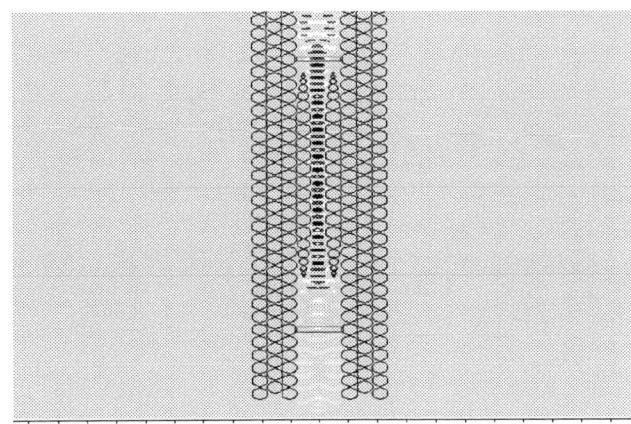

Fig.4 (a) the mode propagation through the proposed PC taper

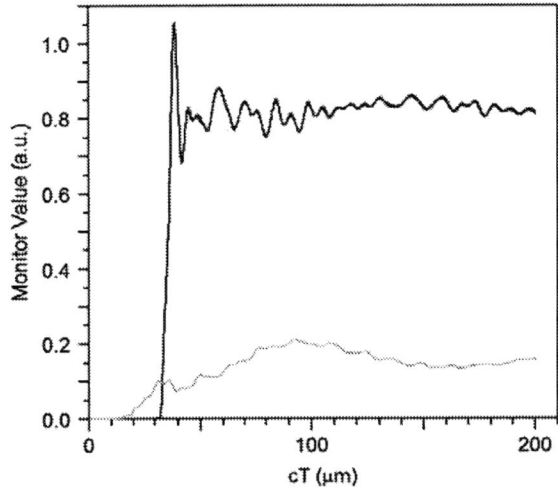

Fig. 4 (b) the transmission efficiency (blue curve) and the reflection efficiency (green curve) of the proposed PC taper

In this work, a novel taper structure which is designed on a two-dimensional PC consisting of a triangular array of holes drilled on a Si confining layer has been proposed and validated through computational simulation. The proposed PC taper combines two typical taper structures together, which can not only reduce the mode field mismatch, but also transform between the forward propagating mode and bi-direction propagating mode. The FDTD simulation results prove that this PC taper has high transmission efficiency of 85% and can convert the mode profile smoothly. It has very great potential in all kind of PC components.

ACKNOWLEDGMENT

The authors thank the members of SPM (Silicon Photonics and Microsystems) group for their helpful discussions.

REFERENCES

[1] A. Talneau, Ph. Lalanne, M. Agio, C. M. Soukoulis, "Low-reflection photonic-crystal taper for efficient coupling between guide sections of arbitrary widths," OPTICS LETTERS, 2002, 27(17), pp.1522-1524

[2] Peter Bienstman, Solomon Assefa, Steven G. Johnson et al. Taper structures for coupling into photonic crystal slab waveguides. Optical Society of America, 2003, 20(9), pp.1817-1821

[3] Ph. Lalanne, A. Talneau. Modal conversion with artificial materials for photonic-crystal waveguides. OPTICS EXPRESS, 2002, 10 (8), pp. 354-359

Tunable Wavelength Multicasting Using Pulsed Pumping Four-Wave Mixing in a Highly Nonlinear Fiber

Jian Wang, Junqiang Sun, Qizhen Sun, Hui Cao

Wuhan National Laboratory for Optoelectronics, School of Optoelectronic Science and Engineering, Huazhong University of Science and Technology, Wuhan 430074, Hubei, P. R. China

Abstract—**We report tunable wavelength multicasting of picosecond pulses based on pulsed pumping four-wave mixing (FWM) in a highly nonlinear fiber (HNLF). By employing two or three input continuous-wave (CW) signals, single-to-dual, single-to-four, and single-to-six channel wavelength conversions are observed in the experiment, taking into account the evolution of CW optical waves into optical pulses during FWM.**

I. INTRODUCTION

Recently, all-optical wavelength conversions using cascaded second-order nonlinearities of periodically poled lithium niobate waveguide (PPLN) and four-wave mixing (FWM) of highly nonlinear fiber (HNLF) have attracted considerable interest [1]. They offer several distinct advantages required by an ideal wavelength converter, such as ultra-fast response, no spontaneous emission noise, complete transparency and independence to bit rate and data format, etc. Previously, we have studied various PPLN-based wavelength conversions of picosecond pulses [2-5], especially including wavelength multicasting, i.e. single-to-dual and single-to-triple channel wavelength conversions by use of cascaded second-harmonic generation and difference frequency generation (SHG+DFG) with pulsed pumping [4, 5]. In this paper, by exploiting pulsed pumping FWM in an HNLF, tunable wavelength multicasting (single-to-dual, single-to-four, and single-to-six) of picosecond pulses is also experimentally demonstrated.

II. OPERATING PRINCIPLE AND EXPERIMENTAL SETUP

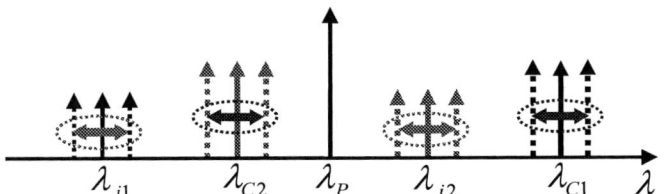

Fig. 1. Operating principle of FWM-based tunable wavelength multicasting.

Fig. 1 shows the operating principle of the proposed scheme. Note that here the information is imposed on the pump wavelength (λ_P). During the degenerate FWM process, two converted idler ($\lambda_{i1}, \lambda_{i2}$) waves will be generated as two continuous-wave (CW) optical sources ($\lambda_{C1}, \lambda_{C2}$) are incident, satisfying the wavelength relationship of $2/\lambda_P = 1/\lambda_{C1} + 1/\lambda_{i1}$ and $2/\lambda_P = 1/\lambda_{C2} + 1/\lambda_{i2}$, respectively. Remarkably, besides the wavelength conversions from the pump wavelength to dual channel idler wavelengths, the input CW optical sources are also modulated by the pulsed pump and will change into optical pulses after FWM, because of which the output λ_{C1} and λ_{C2} will also take the information carried by the pulsed pump. Therefore, single-to-four channel wavelength conversion can be realized by employing two input CW optical sources. Moreover, single-to-dual and single-to-six channel wavelength conversions will be achieved with one or three CW incident optical sources.

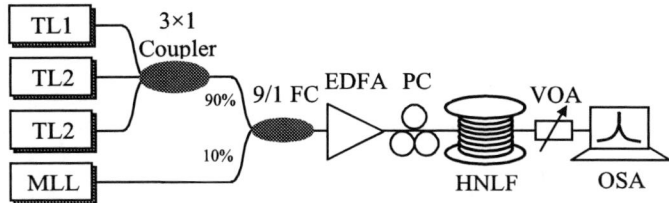

Fig. 2. Experimental setup for FWM-based tunable wavelength multicasting: VOA: variable optical attenuator.

The experimental setup for pulsed pumping FWM-based tunable wavelength multicasting is shown in Fig. 2. A mode-locked fiber laser (MLL) with a repetition rate of 40 GHz and a pulse width of 1.57 ps serves as the pulsed pump with its central wavelength set at 1556.0 nm. Three CW tunable lasers (TLs), combined with the pulsed pump with a 9/1 fiber coupler (FC), are amplified by a high-power erbium-doped fiber amplifier (EDFA) with a small-signal gain of 40 dB and saturation output power of 30 dBm. The amplified optical waves are then launched into an HNLF and participate in the FWM process. The HNLF has a length L=135 m, nonlinear coefficient $\gamma = 20 \, (W \cdot km)^{-1}$, dispersion slope $S = 0.031 ps/nm^2/km$, a fiber loss of 0.51 dB/km, a dispersion zero wavelength at 1552.0 nm, and an effective area of 11 μm^2. The polarization controller (PC) is placed before HNLF to adjust polarization states of incident optical waves and enhance the nonlinear interaction in the HNLF. The output optical spectra are monitored by an optical spectrum analyzer (OSA, Anritsu MS9710C) with the highest spectral resolution of 0.05 nm.

III. EXPERIMENTAL RESULTS AND DISCUSSIONS

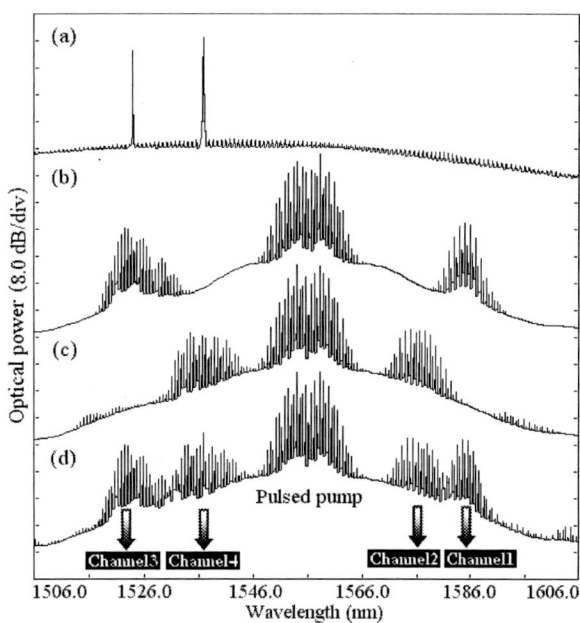

Fig. 3. (a) Two input CW optical waves at 1523.8 and 1536.8 nm. (b)(c) Tunable single-to-dual channel wavelength conversion using only one input CW optical wave. (d) Single-to-four channel

Fig. 3 clearly illustrates the typical output spectra for single-to-four channel wavelength conversion. TL1 and TL2 in Fig. 2 are employed, producing two input CW optical waves at 1523.8 and 1536.8 nm, respectively. The central wavelength of the pulsed pump is set at 1556.0 nm. Because of the FWM in HNLF, two new idler waves at 1585.6 and 1575.2 nm are generated (channels 1 and 2), taking the information carried by the pulsed pump. Note that the input CW optical waves have very simple linear spectra as shown in Fig. 3(a), however, after experiencing FWM, some new sidebands show up in the output spectra, which are similar to the input pulsed pump. Such phenomenon can be explained with the fact that two CW optical waves are modulated by the pulsed pump and therefore evolved into optical pulses at the output of HNLF. Thus the information carried by the pulsed pump is also copied onto two incident wavelengths, corresponding to channels 3 and 4 shown in Fig. 3(d). That is, single-to-four channel wavelength conversion is successfully implemented. In addition, when using only one input CW optical wave, single-to-dual channel wavelength conversion can be achieved as shown in Fig. 3(b)(c). In fact, Fig. 3(b)(c) exhibits tunable single-to-dual channel wavelength conversion for a fixed pump wavelength by simply changing the wavelength of the input CW optical wave.

Fig. 4 depicts the tunable single-to-four channel wavelength conversion by adjusting two input CW optical waves at 1605.2 and 1525.0 nm (channels 3 and 4), respectively. The two corresponding new idler waves are generated at 1505.6 and 1585.3 nm (channels 1 and 2).

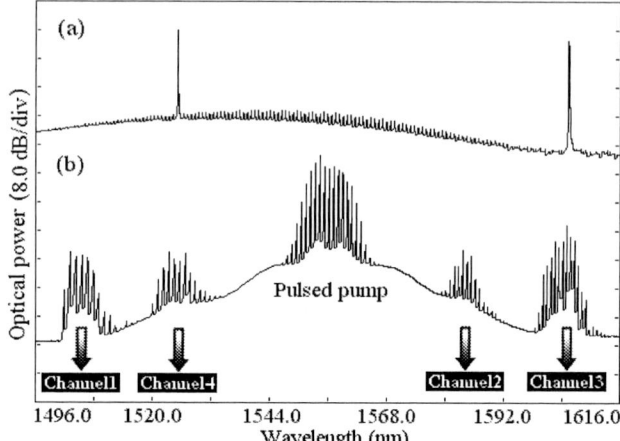

Fig. 4. (a) Two input CW optical waves at 1605.2 and 1525.0 nm. (b) Single-to-four channel wavelength conversion.

Fig. 5 further presents the single-to-six channel wavelength conversion by simultaneously using TL1, TL2, and TL3 shown in Fig. 2. Three input CW optical waves at 1600.4 (channel 4), 1526.2 (channel 5), and 1538.2 nm (channel 6) interact with the pulsed pump at 1556.0 nm to yield three new idler waves at 1510.6 (channel 1), 1584.3 (channel 2), and 1575.0 nm (channel 3), respectively. By comparing the measured optical spectra for six channels with the pulsed pump spectrum, it can be clearly seen that single-to-six channel wavelength conversion is successfully realized. Furthermore, tunable operation can also be performed by changing the wavelength of input CW optical waves.

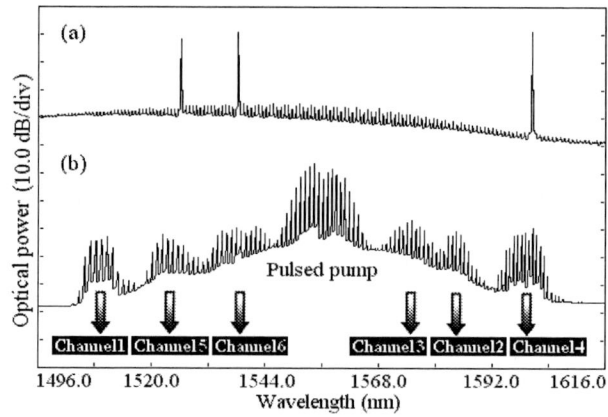

Fig. 5. (a) Three input CW optical waves at 1600.4, 1526.2, and 1538.2 nm. (b) Single-to-six channel wavelength conversion.

Additionally, Figs. 3-5 imply that it is possible to perform tunable single-to-multiple channel wavelength conversion by increasing the number of input CW optical waves.

IV. CONCLUSION

Tunable wavelength multicasting of picosecond pulses based on pulsed pumping FWM in an HNLF is proposed and demonstrated. Single-to-dual, single-to-four, and single-to-six channel wavelength conversions are successfully observed.

ACKNOWLEDGMENT

This work was supported by the National Natural Science Foundation of China under the Grant No. 60577006.

REFERENCES

[1] S. J. B. Yoo, "Wavelength conversion technologies for WDM network applications," J. Lightw. Technol., vol. 14, pp. 955–966, Jun. 1996.

[2] J. Wang, J. Sun, and Q. Sun, "Experimental observation of a 1.5 μm band wavelength conversion and logic NOT gate at 40 Gbit/s based on sum-frequency generation," Opt. Lett. vol. 31, pp. 1711-1713, Jun. 2006.

[3] J. Wang, J. Sun, J. R. Kurz, and M. M. Fejer, "Tunable wavelength conversion of ps-pulses exploiting cascaded sum- and difference frequency generation in a PPLN-fiber ring laser," IEEE Photonics Technol. Lett., vol. 18, pp. 2093-2095, Oct. 2006.

[4] J. Wang, J. Sun, J. Li, and Y. Guo, "Single-to-dual channel wavelength conversion of picosecond pulses using PPLN-based double-ring fibre laser," Electron. Lett., vol. 42, pp. 236-237, Feb. 2006.

[5] J. Wang, J. Sun, C. Luo, Q. Sun, X. Zhang, and D. Huang, "Single-to-multiple channel wavelength conversions and tuning of picosecond pulses in quasi-phase-matched waveguides," Chin. Phys. Lett., vol. 23, pp. 1806-1809, Jul. 2006.

FDTD Simulation of Band Structure and Mode Distribution for Plasmonic Crystals

Bin Zhang[a], Yi Wang[a] and Zhiping Zhou [a,b]

[a]Wuhan National Laboratory for Optoelectronics, Huazhong University of Science and Technology,
Wuhan, Hubei 430074, China
[b]School of Electrical and Computer Engineering, Georgia Institute of Technology,
Atlanta, Georgia 30332, USA

Abstract—**We use Finite-Difference Time-Domain (FDTD) method to calculate surface plasmon modes distribution as well as band structures for the typical two dimensional plasmonic crystals. The unique properties of compact flat bands, highly degenerated surface plasmons are proved to enhance the electric field intensity significantly. The corresponding modes distribution are simulated and discussed.**

I. INTRODUCTION

The surface plasmon polaritons (SPPs), existed at the interface between metal and dielectric [1], are transverse magnetic wave which is shown in Figure 1 (a). By adding or subtracting external momentum, the photons can be coupled into the SPPs. For other structure like prills or holes, localized surface plasmons (LSPs), shown in Figure 1 (b) are the oscillations of electron cloud. The LSPs can be excited directly when the frequency of the incident light is close to the resonant frequency of the LSPs. Because the electrons can be highly localized in very small volume, the field enhancement caused by the LSPs is up to 10^5 [2].

Here we report the simulation results of band structure and mode distribution for the two dimensional plasmonic crystals. Due to strong electron-photon coupling, the compact flat bands

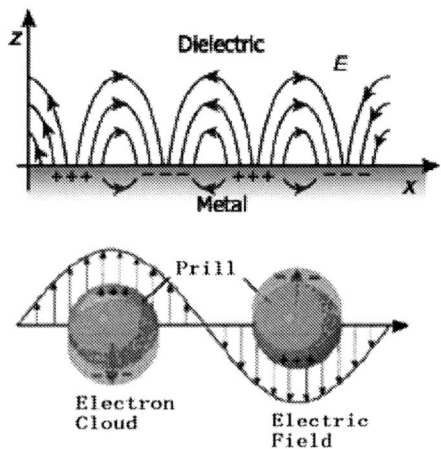

Fig.1 Schematic illustration of the collective oscillations of free electrons for a metal–dielectric interface and metal prill [2].

that reduce the band dispersion [3] are highly degenerate.

II. SIMULATION RESULT

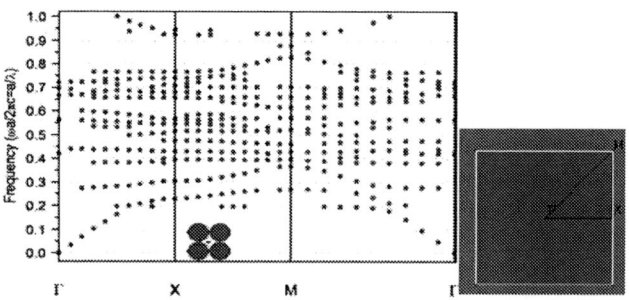

Fig.2 The left picture is calculated band structure of square lattice with circle column which is shown in the downside, the y-axis is frequency which is normalized as $\omega a / 2\pi c = a / \lambda$, the x-axis is wave vector in the first Brillouin zone which is show in the right.

We calculated the band structure and mode distribution by using the FDTD method which is illustrated in Figure 2 and Figure 3, respectively. Here r is the radius of the column, a is the lattice constant and r/a is equal to 0.3. The material of the columns are silver and the dielectric function of silver, which is, defined as the model of the free electron gas, is given as below:

$$\varepsilon(\omega) = 1 - \frac{\omega_p^2}{\omega^2} \qquad (1)$$

where ω_p is the bulk plasma frequency of the silver. All the results are done with the software of Rsoft Corporation.

The asymptotic surface plasma frequency at the large-k limit of the dispersion relation of metallic slab surface is a good approximation for some two dimensional structures [5, 6].And it takes the form as:

$$\omega_{sp} = \frac{\omega_p}{\sqrt{1+\varepsilon_d}} \qquad (2)$$

where ε_d is the dielectric constant. When the metal is surrounded by air, namely $\varepsilon_d \approx 1$, the corresponding ω_{sp} is about $0.7\omega_p$. We find most of the flat bands are around $0.7\omega_p$.

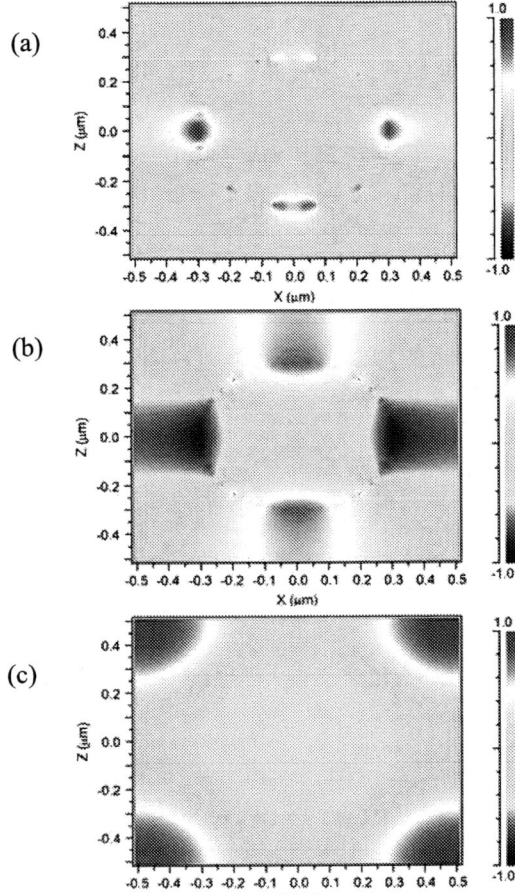

Fig.3 Computed mode distribution when the normalized frequency is (a) 0.681, (b) 0.5 and (c) 1.1.

In Figure 2, some flat bands can be found close to ω_{sp}. The corresponding mode distribution is illustrated in Figure 3. As it is shown in Figure 3 (a), the electric field intensity is strongly localized at the surface of the metallic column when the frequency of the excitation source is close to $0.7\omega_p$. However, in the Figure 3(b), the electric field is localized at the surface of the metal and expands its field distribution into space when the frequency of the excitation source is a little away from $0.7\omega_p$. In Figure 3(c), there is no localized electric field at the surface of the metal and all the electric field is located in the space. In this case, the frequency of excitation source is far away from $0.7\omega_p$.

In Figure 4(a), the band structure of the plasmonic crystal in square lattice with hexangular pillar is shown, whereas the band structure of the plasmonics crystal in square lattice with square pillar is illustrated in Figure 4 (b).Comparing to the Figure 2 (a), the number of flat bands in Figure 4 increase gradually. It's maybe the reason that the hexangular or tetragonal pillars which have corners on surface of metal will gather more charges, and in reverse, they will induce more LSPs at the same frequency.

Furthermore, the flat bands, which are highly degenerate, will result in strongly amplified electric field intensity. At the interface between metal and dielectric there are a lot of eigenmodes corresponding to the flat bands. Comparing Figure 4 (a) with Figure 4 (b), we find that there will be more eigenmodes if the curvature of the corner is larger. Moreover,

the number of the eigenmodes will increase linearly with more calculation grids .This result is the same with that of other papers using the method of atomic orbit functional linear combination [5].

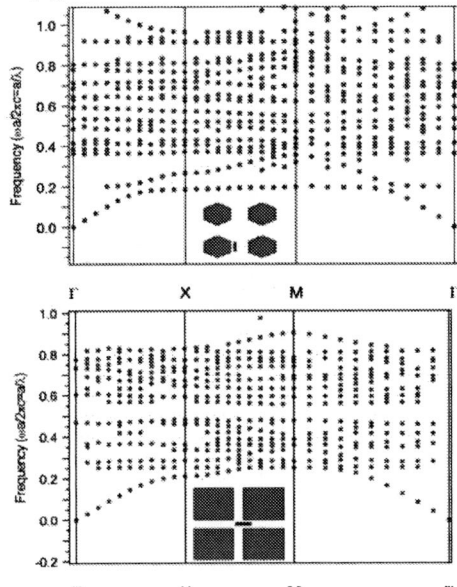

Fig.4 Calculated energy band structure of square lattice with hexangular or square column

Contrary to SPPs, the resonance condition of LSPs exists infinite mode.For example, on a metallic spherical surface, the resonance condition of the LSPs is:

$$\varepsilon_1'(\omega) = -\varepsilon_2 \frac{l+1}{l}, l = 1,2,3...... \tag{3}$$

where l is the angular momentum of the LSPs, $\varepsilon_1'(\omega)$ is real part of metallic dielectric constant, ε_2 is dielectric constant for the environment, ω is frequency of the incident light.

A metallic ellipsoid, and its major axis is a and the minor axis is b. When $a, b \gg \lambda$ and the incident electric field parallel to major axis, the polarization can be represented as

$$P(\omega) = \frac{1}{4\pi} \frac{\varepsilon_1(\omega) - \varepsilon_2}{\varepsilon_2 + (\varepsilon_1(\omega) - \varepsilon_2)A_\alpha} E_0 \tag{4}$$

where Aa is a depolarization factor and it can be definition as

$$A_a = \frac{ab^2}{2} \int_0^\infty \frac{ds}{(s+\alpha^2)R} \tag{5}$$

where $\alpha = a$ or b, $R^2 = (s+a^2)(s+b^2)^2$ When the shape of the particle is sphere, $Aa = 1/3$. When the shape of the particle is spheroid and $b/a = 1/3$, $Aa = 0.1087$. The field amplified coefficient (FAC)is:

$$T = \left| \frac{E_{tip}}{E_0} \right| = \left| \frac{\varepsilon_1}{1+(\varepsilon_1-1)A_\alpha} \right| = \left| \frac{\varepsilon_1'}{\varepsilon_1'' A_\alpha} \right|^2 \tag{6}$$

When depolarization factor Aa decrease, the resonance wavelength will be shifted toward the long wavelength and the enhancement factor increase. For example, when the ratio of major axis to minor axis of metallic spheroid is $b/a = 1/3$, depolarization factor is $Aa = 0.1087$. If the metallic material is silver, the resonance wavelength is 480 nm and $\varepsilon_1^{"}(\omega) = 0.34$. After the calculation from Eq. (6), the enhancement factor at the tip of major axis can achieve as high as 4.9×10^4. By the way of the above analysis and computation, it is clearly understood that the intensity of electric field of the LSPs is very large.

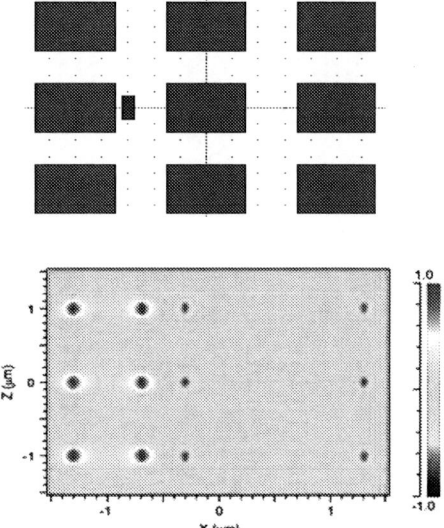

Fig.5 Entire dimensional distribution of mode field

Our structure is very similar to the ellipsoid with infinite length in one dimension ,so we should get the very same FAC. As a first step to get the FAC ,we should get a whole view of mode distribution of a 3×3 plasmonic crystals in square lattice with square column as is illustrated in Figure 5. As it is shown in Figure 5, the electric fields are all strongly localized around the columns. In order to estimate of the electric field intensity caused by the LSPs, We define the FAC as the ratio of the field intensity on the metal surface to the field intensity of the excitation source at very same frequency.

$$T = \frac{F_1}{F_2} \tag{7}$$

where F_1 is electric field intensity at the detector placed at the surface of the metal, and F_2 is the field intensity of excitation source. The general choice of the frequency is ω_{sp} [2].

By simple calculation, the FAC of the plasmonic crystal at ω_{sp} in Figure 6 is given as below:

$$T = 21.482 \times 10^5 / 6.1 \approx 3.5 \times 10^5$$

It means the electric field intensity is amplified over five magnitudes than the intensity of the excitation source at the same frequency.

III. CONCLUSION

In this paper, we calculated the band structure and mode distribution of some typical two dimensional plasmonic crystals. We find the flat band is associated with the corners of the columns or pillars within the plasmonic crystals and the curvature of the corners due to the existence of the LSPs. Because the LSPs are strongly localized at the surface of the silver, the electric field intensity near the silver surface can be amplified distinctly.

ACKNOWLEDGMENT

The authors gratefully acknowledge the valuable advice and technical assistance provided by Prof. Lai.

REFERENCES

[1] T.W.Ebbesen, H.J.Lezec, "Surface plasmons enhance optical transmission through subwavelength holes," *Phys. Rev. B*,58,6779-6782

[2] Heinz Raether, "Surface Plasmons on smooth and rough surfaces and on gratings," Springer,1986

[3] W. Zhang, X. L. An Hu, N. Xu, and N. Ming, "Photonic band structures of a two-dimensional ionic dielectric medium", *Phys. Rev. B 54*,10280 ,1996

[4] Younan Xia and Naomi J. Halas, " Shape-Controlled Synthesis and Surface Plasmonic Properties of Metallic Nanostructures," *MRS BULLETIN • VOLUME 30 • MAY 2005*

[5] Takunori Ito and Kazuaki Sakoda, "Photonic bands of metallic systems. II. Features of surface plasmon polaritons," *Phys. Rev. B 64,045117*

[6] E. Moreno, D. Erni, and C. Hafner, "Band structure computations of metallic photonic crystals with the multiple multipole method,"*Phys. Rev. B 65, 155120*,2002.

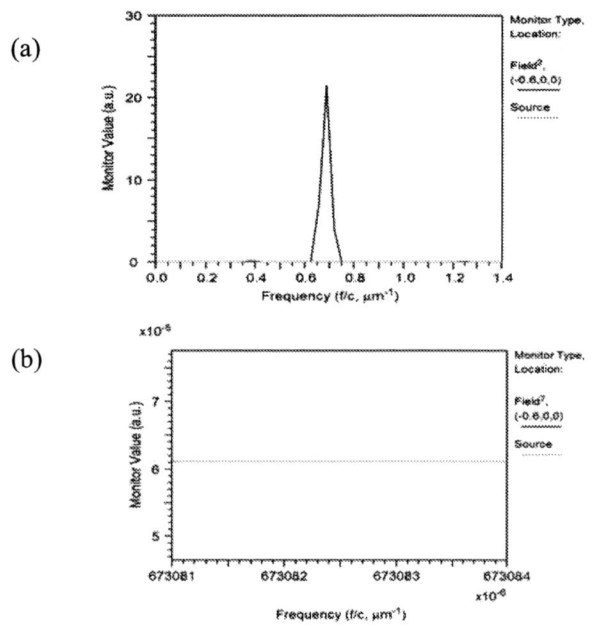

Fig.6 blue line represents the electric field detected by the time

Measurement of Gain Curves for Semiconductor Optical Amplifier Utilizing Hakki-Paoli Method With Wavelet Denoise and Deconvolution Process

Lei Liu, Xinliang Zhang and Dexiu Huang

Wuhan National Laboratory for Optoelectronics and Institute of Optoelectronic Science and Engineering,
Huazhong University of Science and Technology, Wuhan, 430074

Abstract—**For improved the precision of the measurement result of the gain coefficient and avoid the affection of the optical spectrum analyzer's resolution and noise, we combined the Hakki-Paoli method with the wavelet denoise and deconvolution process. It was testified from the simulation that this method improve the precision of the gain coefficient measurement. At last, the conclusion was testified from the experiment.**

I. INTRODUCTION

The optical gain of a semiconductor optical amplifier and semiconductor lasers is an important parameter for device performance and design, especially for semiconductor optical amplifier that focus on nonlinear effect. The techniques for measurement of gain have been investigated intensively for three decades.

By measuring the spontaneous emission spectrum in the vertical or lateral direction of a semiconductor optical amplifier, we can obtain the gain spectrum based on the relation between the gain spectrum and the spontaneous emission spectrum. For the conventional Fabry–Pérot semiconductor lasers, the amplified spontaneous emission modulated by the laser cavity is usually used for measuring the gain spectrum by the Hakki-Paoli (HP) method.

The HP method, which derives the gain from the modulation depth of the ASE spectrum, is sensitive to the resolution of the measurement system and may underestimate the gain greatly as the ripple is not obviously. owing to the difficulty to obtain the true peak above threshold and to the large influence from the instrument response function. Cassidy modified the Hakki–Paoli (H–P) method by using the ratio of the integral (mode sum) of the wavelength resolved power over one mode to the minimum power instead of the peak-to-valley ratio to derive the gain. This mode sum/min method is much less sensitive to the response function of the optical spectrum analyzer (OSA) than the H–P method, and thus can be extended to measurements above threshold. A de-convolution technique was reported by Cassidy to minimize the effect of the instrument response function. Considering the OSA as a linear response system, the measured ASE spectrum is the convolution of an intrinsic laser ASE spectrum with the OSA response function. Such a convolution

process has an intermixing effect between adjacent ASE peaks, which causes the HP method to underestimate the gain. Meanwhile, Both the H–P and mode sum/min methods are sensitive to the noise as a result of the small intensity and a single measurement at the minimum. Although the noise can be reduced by averaging in the region around the minima, the effect of the averaging on the value obtained for the minima must be considered. Recently, the Fourier transform (FT) method and Fourier series expansion (FSE) method was proposed. Consistent results with the HP method were obtained by the FT method. Expect the convenience, FT method Concerned weighted integrals over the mode and not a single point measurement so that the noise is partly reduced by summation but FT method still limited by the resolution of the OSA. Meanwhile, for overcome the limit resolution of the OSA, FT with a de-convolution process (FTD) was proposed to calculate the gain spectra from the measured ASE spectrum. But the FTD method is more sensitive to noise than FT method because of the de-convolution.

In this paper, we proposed a new method ---HP-DD to overcome the effect of the noise and the resolution of the OSA.HP-DD is a modified version of the Hakki-Paoli method.In this paper, we first introduce the Hakki-Paoli method, the wavelet denoise and the de-convolution process in Section II, and then present the numerical simulation results with wavelet de-noise and de-convolution process based on ASE spectra convoluted with response function of OSA in Section III. The numerical results obtained by the HP-DD and Hakki-Paoli methods are compared with the practical values. The measured gain spectrum for a SOA are presented in Section IV. Finally, we get the conclusion in Section V.

II. HP-DD METHOD

The gain coefficient spectrum has essential relativity with the emit spectrum. The material gain coefficient spectrum can be measurement by SPE spectrum. Hakki and Paoli proposed a method that measuring the model-gain from the modulation depth of the amplified spontaneous emission spectrum. It is based on the formula:

1-4244-0816-4/06/$25.00 ©2006 IEEE

$$g_{net} = 1/L * \ln(\frac{r_i^{1/2} - 1}{r_i^{1/2} + 1}) + 1/L * \ln(1/R) \qquad (1)$$

R is the reflectivity of the SOA. But the accuracy of measured gain spectra is usually limited by the resolution of the optical spectrum analyzer (OSA), a deconvolution process based on the measured spectrum of semiconductor optical amplifier is applied to improve the accuracy in the Hakki-Paoli method. Generally, the measurement process of the OSA can be express as:[6]

$$h * r + n = f \qquad (2)$$

h is the response function of the OSA in different resolution, r is the real ASE spectrum, n is the noise in the measuring process, f is the observed spectrum. The response function h can be written as its convolution matrix H, so the formula (2) can be rewrited as

$$Hr + n = f \qquad (3)$$

The FTD method realized the de-convolution process in the cavity length field. In this paper, we realize it in the spectrum directly. As the h function is low-pass, so the de-convolution process is abnormal. In this paper, we choose a new algorithm to realize the de-comvolution:

$$r_{k+1} = P \cdot r_k \qquad (k = 0,1,2...) \qquad (4)$$

where $r_0 = f$, $r' = P \cdot r$ is equalized to

$$r' = r + (H'H + 1/\lambda * I)^{-1} H'(f - Hr) \qquad (5)$$

This algorithm makes the deconvolution of the low passfunction to normal,meanwhile the measuring process is sensitive to the noise because of the de-convolution.If we denoise the signal before de-convolution, the result will be more accurate. The wavelet denoise can process the signal in temporary and frequency field simultaneously in multi frequency scale.So we used the wavelet denoise in this paper.The accuracy can be improve after the denoise and deconvolution which is the base of our method.

III. NUMERICAL SIMULATION

Based on the transmission line formalism,the ASE of the SOA can be described by the function:

$$I(\beta) = A\frac{(1-R)^2}{R}\frac{b}{1+b^2-2b\cos(4\pi\beta nL)} \qquad (6)$$

where $b(\beta) = \exp[g(\beta)L]R$, β is the wavenumber, A is a slowly varying function of the wavenumber, R is the reflectivity of the SOA, nL is the product of the refractivity and cavity length.The function can be seen as the r in the formula (3).For simulate the real measuring process, we convoluted the r with the response function in different resolution which get from the OSA directly. The convoluted spectrum can be seen as the $h * r$ .Then we can add the Gaussian distribution white noise to $h * r$ for stimulate the noise in the measurement and the essential noise in the ASE. Now we present the ASE with noise is f in the $Hr + n = f$. The real r and f in different resolution is what we analyzer and compare in the follow simulation.

Then we implement the HP-DD on the f .At first, we process the ASE with noise by the wavelet denoise. We choose the Harr5 wavelet function to decompose the ASE with noise in five level wavelet coefficients. What we must pay attention is that the ripple in ASE spectra can be seen as the noise in a way, but its weight focus on the high-level coefficient. The noise in the measuring is focus the low-level coeffient.So we denoise the ASE in the low-level mainly. We choose the threshold denoise with soft threshold.

For validate the affection of the noise on the measurement. We note the RMS of the D1-level coefficient of the real ASE spectra as the reference Rv. Then we add the noise which has different proportion of the Rv to the convoluted ASE for simulate the different measured ASE. Then we obtained gain coefficient spectra from the ASE without and with the noise utilizing the Hakki-Paoli method. The relatively error was plotted in figure1 by diamond and empty dot. We can see that the error is increasing with the percentage of the error. On the other hand, we process the ASE with different percentage of noise with the wavelet denoise then we obtained the gain coefficient spectra from the denoised ASE and real ASE utilizing the Hakki-Paoli method. The relatively error was plotted in figure1 by the black dot.We can see that the measurement accurate has been improved greatly from the comparison of the figure1.

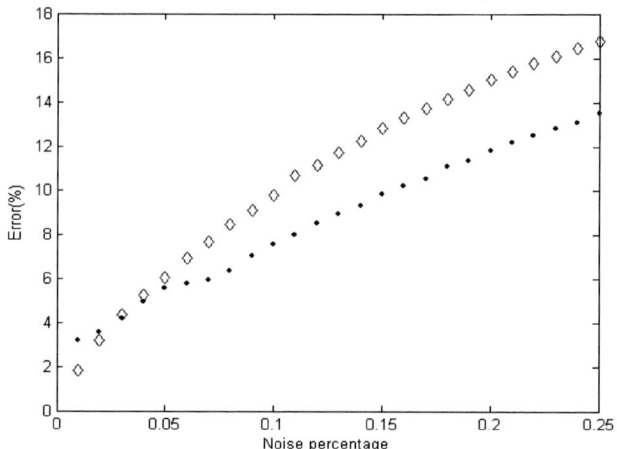

Fig1 the relative error between the gain coefficient obtained from the ASE with noise and after denoise and that from the real ASE

We process the ASE spectra after denoise with the deconvolution process in HP-DD method. The response function can be get from the spectra of the DFB laser in corresponded resolution. The response function of the OSA when resolution be set to 0.5nm is plot in figure2.

We obtained the ASE spectra from the real ASE and the convoluted ASE in different resolution by Hakki-Paoli method and plotted it in fig3.We can see the affection of the resolution clearly .

In fig3,the top one is the result of the real ASE spectrum. Then it is the results of the ASE spectra when resolution is 0.05nm, 0.1nm and 0.5nm from top down. We can see that the bigger the resolution, the more blur the ASE spectra. So the result is smaller.Then we obtain the results from the ASE after

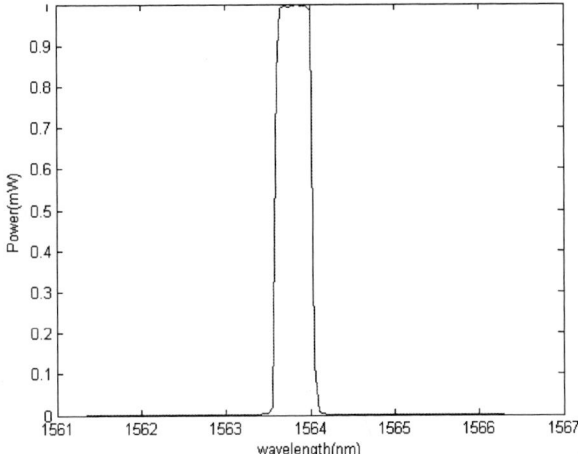

Fig2.the response function of the OSA when resolution is 0.5nm

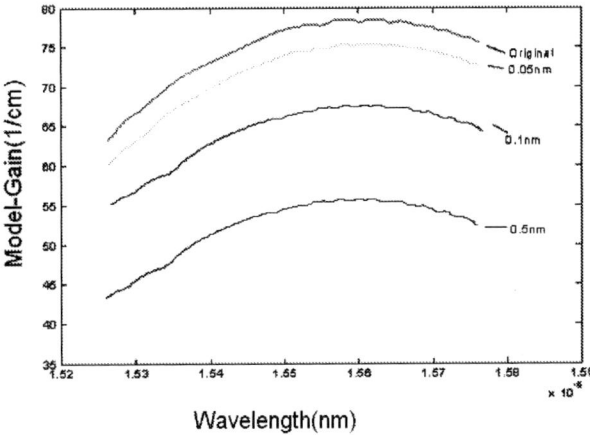

Fig3 The gain coefficient spectra obtained from the real ASE and convoluted ASE in different resolution.

deconvolution in different resolution and compare it with the result in fig3.Note the result of the real ASE is the standard, we can get the relative error of the resolution with and without the deconvolution process. The result after deconvolution close to a fixed value and the result are improved greatly.

IV. EXPERIMENT RESULT

In the simulation, we have seen the validity of the HP-DD method. We will testify that in the measurement. A single-port semiconductor optical amplifier with 1mm cavity length, the total reflectivity about 10-4 and inject current 230mA is used for the measuring the gain spectrum. The ASE spectrum of the SOA is measured with the Antrisu MS9710C OSA which the resolution set to 0.05nm, 0.1nm, 0.2nm and 0.5nm..The response function of the OSA in different resolution is obtained by measuring the DFB tunable laser New Focus6300 at the wavelength of 1564nm.The line width of the laser can be ignored in comparison with the resolution of the OSA.We can see the response function of the OSA at the resolution of 0.5nm in Fig.2 .

Take the measurement result at the resolution of 0.5nm with HP-DD for example. At first, we process the ASE spectrum at the resolution of 0.5nm with the five level wavelet denoise by

Harr5 function. In low level coefficient (detail 1 coefficient),we use the soft threshold denoise. Then we process the deconvolution. The result is plotted in fig4.The top curve is the original ASE obtained at the resolution of 0.5nm,the bottom one is the one after process, it is similar to the ASE obtained at the resolution of 0.05nm. Then we obtain the gain coefficient from the two ASE spectrum with the Hakki-Paoli method. And then we can get the gain coefficient difference using the one original minus the one after process. The result is shown in fig5,the gain coefficient obtain from the ASE spectrum after process is bigger than that form the ASE without process. So the underestimation of the gain of the Hakki-Paoli method is eliminate in HP-DD method.

TABLE I
THE RELATIVE ERROR IN DIFFERENT RESOLUTION WITH AND WITHOUT DECONVOLUTION

RESOLUTION (NM)	ERROR WITHOUT DECONVOLUTION (1/CM)	ERROR WITH CONVOLUTION (1/CM)
0.05	3.1	0.1
0.1	10.1	0.2
0.5	20.1	0.5

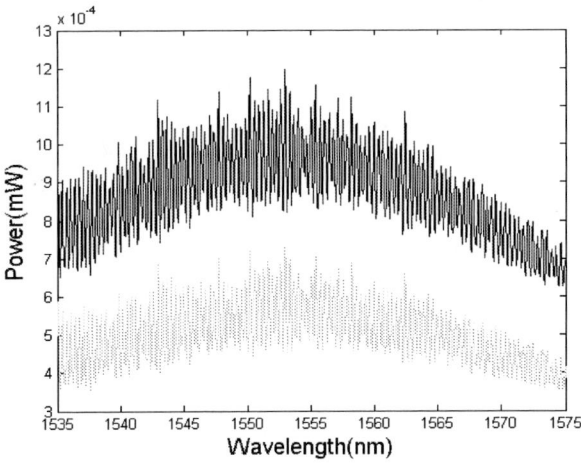

Fig 4 the ASE spectrum before and after de-convolution.

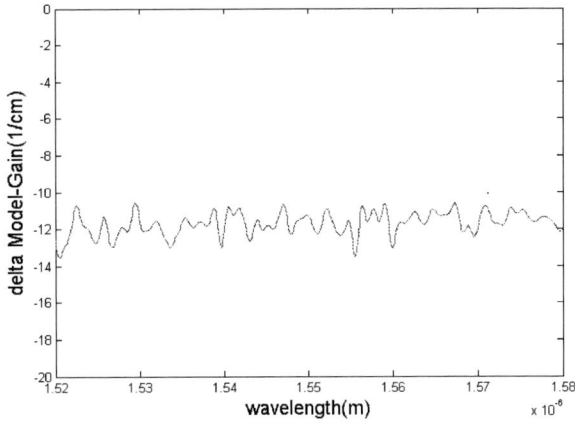

Fig5 The gain coefficient difference obtain before and after deconvolution

Then we can realize the HP-DD method and the Hakki-Paoli method at the resolution of 0.1nm, 0.2nm and 0.5nm.The comparison ware show in Fig6.The top one are the gain coefficient spectra obtained by HP-DD method at different resolution. The following from top down are the gain coefficient spectra obtained by Hakki-Paoli method at different resolution. We can see that result obtained by HP-DD is not sensitive to the resolution and have less ripple even after the deconvolution. The result of the Hakki-Paoli has been improved greatly.

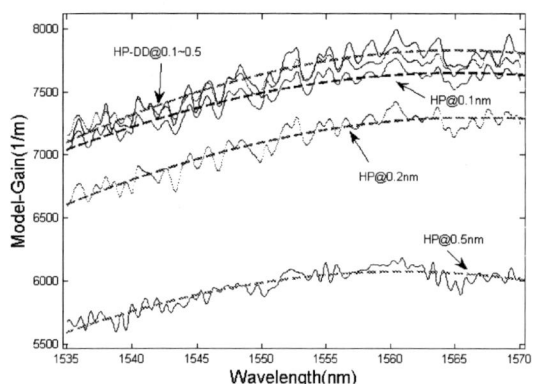

Fig6 The gain coefficient spectra obtained by HP-DD method and the Hakki-Paoli method at different resolution

V. CONCLUSION

In order to eliminate the influence of the limited resolution of the optical spectrum analyzer on the measured gain spectrum and the noise in the measuring process. We proposed a HP-DD method which is a improved method based on Hakki-Paoli method utilizing the denoise and deconvolution process. The response function is obtained from the measuring the spectrum of the DFB laser. The denoise process using the wavelet denoise method. The numerical simulation and the measurement show the great improvement of the method.

REFERENCES

[1] B.W. Hakki and T. L. Paoli, "CW degradation at 300 K of GaAs doubleheterostructure junction lasers, II. Electronic gain," *J. Appl. Phys.*, vol.44, no. 9, pp. 4113–4119, 1973.

[2] B.W. Hakki and T. L. Paoli, "Gain spectra in GaAs double-heterostructure injection lasers," *J.Appl. Phys.*, vol. 46, no. 3, pp. 1299–1305, 1975.

[3] D. T. Cassidy, "Technique for measurement of the gain spectra of semiconductor diode lasers," *J. Appl. Phys.*, vol. 56, no. 11, pp. 3096–3099,1984.

[4] D. Hofstetter and R. L. Thornton, "Loss measurements on semiconductor lasers by Fourier analysis of the emission spectra," *Appl. Phys.Lett.*, vol. 72, no. 4, pp. 404–406, 1998.

[5] D. Hofstetter and R. L. Thornton, "Measurement of optical cavity properties in semiconductor lasers by Fourier analysis of the emission spectrum," *IEEE J. Quantum Electron.*, vol. 34, pp. 1914–1923, Oct. 1998.

[6] Wei-Hua Guo and Yong-Zhen Huang,"Measurement of Gain Spectrum for Fabry–Pérot Semiconductor Lasers by the Fourier TransformMethod With a Deconvolution Process," *IEEE J. Quantum Electron.*, vol.39, pp.719-721, June 2003.

[7] Connelly, M.J., "Wideband semiconductor optical amplifier steady-state numerical Model", *Quantum Electronics, IEEE Journal of*, 2001. 37(3): p. 439-447

[8] C. H. Henry, R. A. Logan, and F. R. Merritt, "Measurement of gain and absorption spectra in AlGaAs buried heterostructure lasers," J. Appl.Phys., vol. 51, pp. 3042–3050, 1980.

[9] P. Blood, A. Kucharska, J. Jacobs, and K. Griffiths, "Measurement and calculation of spontaneous recombination current and optical gain in GaAsAl-GaAs quantum well structures," J. Appl. Phys., vol. 70, pp.1144–1156, 1991.

[10] XiaoYue,CuiYiping,A new kind of wavelet-based method for spectrum de-convolution. *Journal of Southeast University.*, vol.19, no.1, pp12-13, 2003.

[11] D. Hofstetter and J. Faist, "Measurement of semiconductor laser gain and dispersion curves utilizing Fourier transforms of the emission spectra," *IEEE Photon. Technol. Lett.*, vol. 11, no. 11, pp. 1372–1374,1999.

OFDM-ROF System and Performance Analysis of Signal Transmission

Linghui RAO, Xiaoqiang SUN, Wei LI, Dexiu HUANG
Institute of Optoelectronics Science and Engineering, Wuhan National Lab for Optoelectronics,
Huazhong University of Science and Technology, Wuhan 430074

Abstract—**In this paper, an OFDM-ROF communication system is designed and simulated. Special attention has been paid on chromatic dispersion and nonlinear effect brought by fiber transmission. The simulation results of the fiber transmission part are obtained based on Spilt-step Fourier Method and nonlinear Schrödinger Equation. The theory of pulse compression is utilized which would greatly improve the system performance.**

I. INTRODUCTION

The Radio-Over-Fiber technology transforms the high RF (Radio Frequency) signal to optical signal. In a ROF system, most of the signal processing processes (including coding, multiplexing, and RF generation and modulation) are carried out by CS (Central Station), which makes the BS (Base Station) cost-effective. So, ROF will become a key technology in the next generation mobile communication system [1] [2].The OFDM (Orthogonal Frequency-Division Multiplexing) technology, at the same time, is showing its advantages in speeding up the signal processing; it is also being focused in 4G wireless communication system.

In this paper, we will build up a ROF system based on the OFDM technology. The paper is organized as follows, the second part is the performance analysis of fiber transmission in OFDM-ROF System, the third part is analysis of simulation results, and at last, a conclusion is made.

II. PERFORMANCE ANALYSIS OF FIBER TRANSMISSION IN OFDM-ROF SYSTEM

A. Simulation Model of OFDM-ROF System

ROF realize the transparent transform between RF signal and optical signal. Fig.1 is the scheme of the whole system. The input random binary digits are modulated by OFDM technology, transformed to the digital signal, and then modulated by EAM (Electronic Absorption Modulator) to the 1550nm optical carrier generated by DFB laser. The optical singles transmitted in the optical fibers, and are turned back into the microwave through O/E transform in the receiver, then binary digits can be obtained through OFDM-demodulation[3][4]. For the OFDM modulation part, IFFT enables the parallel transmission of the signal. We add training series for channel evaluation, and rotative prefixion is to diminish the interruption between the signals.

B. Performance Analysis of Optical Pulse Transmission in the Fiber

In this part, we will focus on the transmission performance analysis in the optical fibers. The wave forms of the optical pulses are influenced by attenuation, dispersion and nonlinear effect. In the ROF-OFDM system, because of the short distance of fiber, dispersion and nonlinear effect are mainly concerned [5]. The Split-step Fourier method is adopted for simulating dispersions and nonlinear effect in signal transmission[6].

Gaussian pulses generated by laser propagate in the optical fibers along the axis z; the variety of the intensity can be described by Schrödinger Equation,

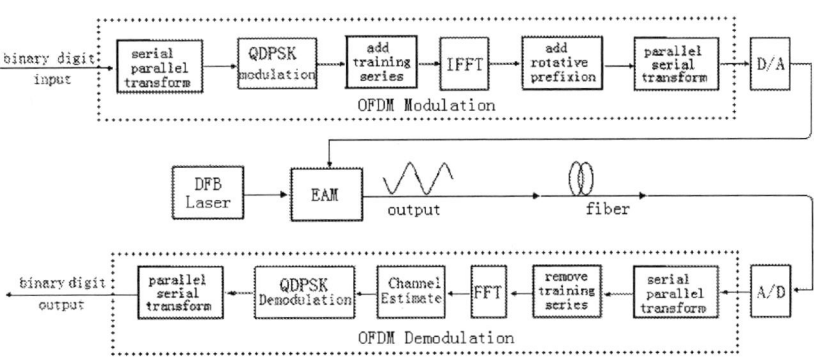

Fig.1. Simulation Model for OFDM-ROF System

1-4244-0816-4/06/$25.00 ©2006 IEEE

$$\frac{\partial A}{\partial z} + \frac{\alpha}{2}A + \beta_1 \frac{\partial A}{\partial t} + \frac{i}{2}\beta_2 \frac{\partial^2 A}{\partial t^2} - \frac{1}{6}\beta_3 \frac{\partial^3 A}{\partial t^3} = i\gamma |A|^2 A \quad (1)$$

where, i denotes the imaginary unit, $A(z,T)$ denotes the amplitude of optical pulses, z is the fiber length, T is the time parameter based on the reference system of group velocity of the central wavelength. $\gamma = n_2 w_0/cA_{eff}$ is the nonlinear coefficient; A_{eff} is the effective mode area. In the following analysis, assume $A(z,T)$ is the normalized amplitude.

According to the Split-step Fourier Method, we obtained

$$\frac{\partial A}{\partial T} = (\hat{D} + \hat{N})A \quad (2)$$

Where, $\hat{D}$ is the differential functor that represents the dispersion and absorption of linear medium, and $\hat{N}$ is the nonlinear functor:

$$\hat{D} = -\frac{i}{2}\beta_2 \frac{\partial}{\partial T^2} + \frac{1}{6}\beta_3 \frac{\partial^3}{\partial T^3} - \frac{\alpha}{2} \quad (3)$$

$$\hat{N} = i\gamma |A|^2 \quad (4)$$

As shown in Fig.2, the Split-step Fourier Method divides the fiber to many parts. Assume the total length of the fiber is L, for each part, the length is h. The dispersion and nonlinear effect can be considered separately in the small part. That is, from z to z+h, first to consider the nonlinear effect, , and ignore the dispersion and loss, so in (2), $\hat{D} = 0$; then, consider the dispersion and loss only, which means $\hat{N} = 0$ in (2). So we obtain:

$$A(z+h,T) \approx \exp(h\hat{D})\exp(h\hat{N})A(z,T) \quad (5)$$

For $\exp(h\hat{D})$ in the Fourier domain:

$$\exp(h\hat{D})U(z,T) = F_T^{-1}\exp[h\hat{D}(i\omega)]A(z,T)F_T\{U(z,T)\} \quad (6)$$

Where F_T is Fourier transform; $\hat{D}(i\omega)$ is obtained from (4) that represents the differential coefficient $\partial/\partial T$ with $i\omega$, ω is the frequency in the Fourier Domain. With the Fast Fourier Transform Method, it is faster to solve (6).

C. Utilization of the dispersion compensation and pulse compression theory

If the theory of pulse compression can be utilized here to make the Gaussian pulse compressed, the performance of the fiber transmission channel will be greatly enhanced.

Pulse compression can be realized by the fiber linking of SMF, DCF, and DSF at a certain ratio.

III. ANALYSIS OF SIMULATION RESULTS

Based on the above model and numerical analysis method, the data transmission process can be quickly simulated by matrixes operation.

For the OFDM coding/decoding process, the simulation shows the accuracy theoretically.

The following analysis is mainly focused on the fiber transmission part.

If DCF (Dispersion Compensation Fiber) is designed for the SMF (Single Mode Fiber) transmission, the negative influence brought by chromatic dispersion can be greatly reduced.

The parameters of SMF and DCF used in the simulation are shown in Table 1. Here, α(dB/km) is the attenuation coefficient;

TABLE I
PARAMETERS OF SMF AND DCF

	α	β2	β3	γ	L1	L2
SMF	0.2	-20	0.08	3	2	4
DCF	0.2	36	0.08	4.8	1.1	2.3

β_2(ps2/km) is the second order dispersion coefficient; β_3(ps3/km) is the third order dispersion coefficient; γ(W^{-1}km^{-1}) is the nonlinear coefficient; L1 is the fiber length of Group 1; L2 is the fiber length of Group 2.

The simulation results are shown in Fig.3 and Fig.4; there are

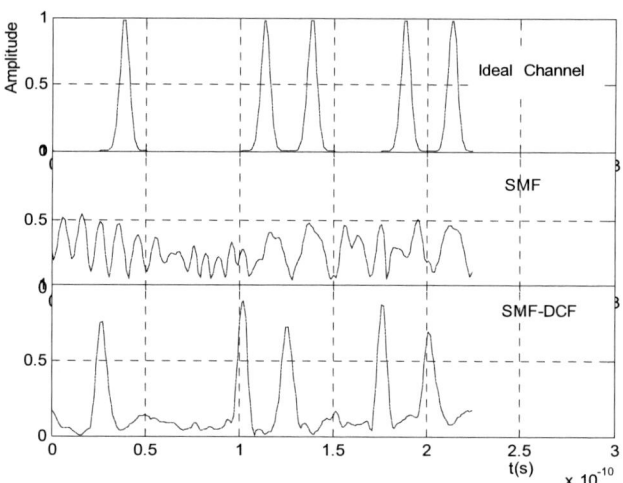

Fig.3. Waveform of different Channels (Group 1)

different pulse-forms after their transmission in different channels. In Group 1, 1.1km of DCF is added after 2.0km of SMF; in Group 2, 2.3km of DCF is added after 4km of SMF.

In Fig.3 and Fig.4, they each have 3 layers, the pulse-form of the top layer is obtained through ideal channel transmission, and there is no distortion. In the middle layer, however, for the transmission through single mode fiber, the pulse-form is already

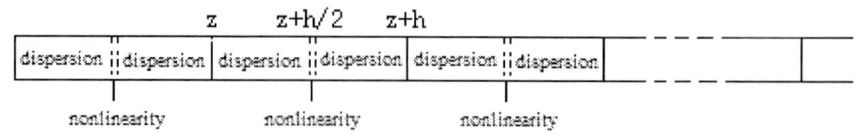

Fig.2. Spilt Fourier Method Model

distorted. At last, the third layer, with DCF compensation, the pulse-form is reshaped.

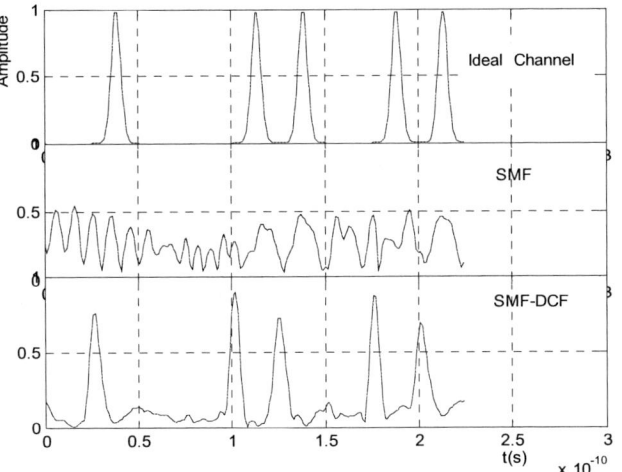

Fig. 4 Waveform of different Channels (Group 2)

In Fig.3, the length of SMF is 2km, and the pulse-form is distorted although it can be identified; when transmitted with the 1.1km DCF compensation, the stretched Gauss pulse can be reshaped. But when the single mode fiber is 4km in length, as shown in Fig.4, the wave-form is no longer recognizable in the first layer, with the 2.3km DCF compensation, the performance is improved. Compared with Group 1(Fig.3), Group 2(Fig.4) has much better performance. The reason is that nonlinear effect brings deeper influence that cannot be ignored as the fiber length increases. Another point is that different ratio of different fibers (SMF to DCF) can result in different performances.

The results lead to the prototype contriving of pulse compression theory utilization to OFDM-ROF system. In this way, pulse-stretching by chromatic dispersion and nonlinear effect can be largely avoided. Meanwhile, O/E transformation precision is enhanced. For example, three stages compressing technique can be employed here, which include 1) compensation by normal dispersion fiber, 2) compressing by normal dispersion fiber followed by anomalous dispersion fiber, 3) compressing by CDPF (comb like dispersion profile fiber).

For the first stage, the initial negative chirp of the pulses can be balanced out with nonlinearity induced by positive GVD; for the second stage, the pulses are compressed by SPM (Self-Phrase Modulation) and GVD (Group Velocity Dispersion); for the third stage, constructing CDPF by splicing alternatively the fibers with low nonlinear coefficients and high dispersion and fibers with high nonlinear coefficients and low

dispersion, nonlinearity, such as SPM, and GVD can be greatly restrained [7].

Table II is the parameters of SMF and DSF in the simulation. Where, α(dB/km) is the attenuation coefficient; β_2(ps2/km) is the second order dispersion coefficient; β_3(ps3/km) is the third order dispersion coefficient; γ(W-1km-1) is the nonlinear coefficient; N is the compression stage; L is the fiber length.

As shown in Fig.5 and Fig.6, 1ps pulses can be obtained from the initial 45.9ps pulses in the transmission. Pulses in Fig.5 are compressed by the special fiber splicing in the ROF system. The negative influence brought by dispersion can be controlled. It not only leads to larger capacity by RF signal, but also to the convenience of signal's reshaping.

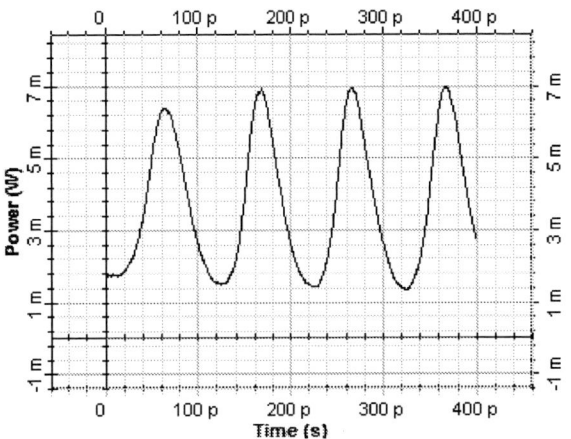

Fig.5. Original waveform

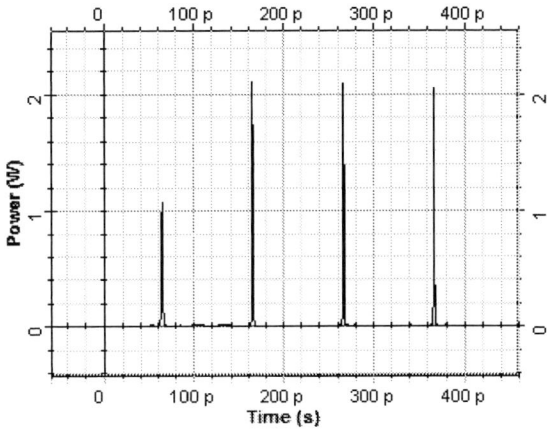

Fig.6. Compressed waveform

IV. CONCLUSION

With the rapid development of mobile communication system, ROF technology, because of its large capacity and lower cost, is becoming a key technology in the future communication system. What's more, the combination of OFDM technology and ROF technology makes it possible to transform the high speed RF signal to the optical signal, so that the optical fibers with broad bandwidth can be used.

TABLE II
PARAMETERS OF SMF AND DSF

	A	B₂	B₃	Γ	N	L(KM)
DCF	0.2	36	0.08	4.8	1	4.2468
DCF	0.2	36	0.08	4.8	2	0.9046
SMF	0.2	-20	0.08	3	2	2.6863
SMF	0.23	-20	0.08	1.62	3	2.4
DSF	0.23	-0.38	0.08	2.36	3	

The simulation of OFDM-ROF system in this paper clearly shows the data processing (coding & decoding) and consequently brings forth the great advantages of OFDM-ROF system. According to the analysis of chromatic dispersion and nonlinear effect of the channels, the utilization of the theory of pulse compression is also proposed and verified by computer implementation here, which would greatly improve the system performance.

REFERENCES

[1] Xiupu Zhang, Baozhu Liu, Jianping Yao, Ke Wu, and Raman Kashyap, "A novel millimeter-wave-band radio-over-fiber system with dense wavelength-division multiplexing bus architecture," *IEEE transactions on microwave theory and techniques*,vol.54,No.2,pp.922-937,Feb.2006.

[2] Toshiaki Kuri, Hiroyuki Toda, and Ken-ichi Kitayama, "Dense wavelength-division multiplexing millimeter-wave-band radio-on-Fiber signal transmission with photonic downconversion," *Journal of lightwave technology*, vol.21, NO.6, pp.1510-1517, June.2003

[3] T.Kuri, K.Kitayama, A.Stöhr, and Y.Ogawa, "Fiber-optic millimeter-wave downlink system using 60GHz-band external modulation," *Journal of lightwave technology*,vol.17,pp.799-806,May 1999.

[4] T.Kuri and K.Kitayama, "Novel photonic downconversion technique with optical frequency shifter for millimeter-wave-band radio-on-fiber systems," *IEEE Photon.Technol. Lett.* vol.14, no.8, pp.1163–1165, Aug.2002.

[5] Senfar Wen and Tsung-Kun Lin, "Ultralong lightwave systems with incomplete dispersion compensations," *Journal of lightwave technology*, vol.19, No.4, pp.471-477, April 2001

[6] G. P. Agrawal. Nonlinear Fiber Optics and Application of Nonlinear Fiber Optics [M]. Third edition. Jia Dongfang, Yu Zhenhong, Tan Bin translated, Beijing: Publishing House of Electronics Industry, 2002:33-35

[7] Hiroyuki Toda, Yasushi Furukawa, Takashi Kinoshita, et al.. Optical Soliton Transmission Experiment in a Comb-Like Dispersion Profiled Fiber Loop [J]. *IEEE Photonics Technology Letters*, vol.9, No.10, pp.1415-1417, 1997.

Author Index

A

Ahn, D. 1
Apollonov, V.V. 26

B

Beals, M. 1

C

Cao, Hui 57
Chen, Jinlin 8
Chen, Yao 8
Chi, Nan 31

D

Dong, Jianji 37

F

Feng, Junbo 8
Fu, Songnian 37

G

Gao, Dingshan 8, 54

H

Hess, Dennis W. 12
Hong, C. Y. 1
Hu, X. 5
Huang, Dexiu 31, 37, 41, 47, 51, 63, 67

J

Jongthammanurak, S. 1

K

Kashyap, Raman 19
Kimerling, L. C. 1
Kovsh, A.R. 5

L

Li, Fang 33
Li, Wei 67
Liu, Deming 41
Liu, J. F. 1
Liu, Lei 63
Liu, S. 5
Liu, Yuliang 33
Liua, Jing 54

M

Michel, J. 1

P

Pan, D. 1
Penty, R.V. 5

R

Rao, Linghui 67

S

Sellin, R.L. 5
Shum, P. 37
Sun, Junqiang 57
Sun, Qizhen 57
Sun, Xiaoqiang 67

T

Thompson, M.G. 5

W

Wang, Jian 57
Wang, Yi 8, 60
Wang, Yongjie 33
Wang, Zhuoran 31
White, I.H. 5
Williams, K.A. 5
Wong, C. P. 12
Wu, Ke 19

X

Xia, Zhixuan 8, 44
Xiao, Hao 33
Xu, Jing 41

Y

Yao, Jianping 19
Yu, Siyuan 31
Yu, Yu 47

Z

Zhang, Bin 60
Zhang, Xiaobei 51
Zhang, Xinliang 37, 41, 47, 51, 63
Zhang, Xiupu 19
Zhou, Zhiping 54, 60
Zhoua, Zhiping 8, 44
Zhu, Lingbo 12

A-1

9781424408160